Editors
Andrew Cherry
Association of Commonwealth Universities
London, UK

James Haselip
DTU - Dept. Management Engineering
UNEP DTU Partnership
Copenhagen, Denmark

Gerard Ralphs
Human Sciences Research Council
Cape Town, South Africa

Isabella E. Wagner
Centre for Social Innovation
Vienna, Austria

https://doi.org/10.1007/978-3-319-69929-5

Preface

The Portuguese capital of Lisbon played host to a historic meeting in the European winter of 2007. It was there that Heads of State and Governments from Africa and the European Union (EU) gathered to agree a new pact—the Joint Africa–EU Strategy (JAES). Setting JAES apart from previous political agreements between the regions has been the explicit inclusion of science and technology (S&T), initially as a distinct chapter of the rolling JAES action plans alongside information society and space, and latterly as a cross-cutting domain. That inclusion of S&T in JAES in part reflected the global consensus at that time that capacity in S&T was essential to economic competitiveness, sustainable development and poverty reduction.

Conceived against this background, the CAAST-Net project launched very soon after the Lisbon Summit. Its purpose? To foster improved quality and quantity of bi-regional cooperation in S&T between Europe and Africa. Targeting areas of mutual interest and benefit, the project gave attention equally to fostering bi-regional partnerships on the one hand through, for example, greater use of EU's Seventh Framework Programme (FP7) funding programme, and on the other hand to bettering the conditions of collaboration, for example, through lobbying for greater coordination between national, and regional research and development policies and instruments.

The 2008 CAAST-Net kick-off meeting in Entebbe, Uganda, brought together the network's then 18 partner organisations, mostly national science authorities from across Africa and Europe and marked the beginning of what was itself to become a long-term Africa–Europe partnership forging new working relationships, exploring new ideas and striking new friendships.

After five years, CAAST-Net gave way at the end of 2012 to CAAST-Net Plus with an expanded consortium of 25 partners, still mostly national science authorities, collectively pursuing the same purpose particularly in relation to the global societal challenges of climate change, food security and health. More closely aligned to the interests and needs of the formal Africa–EU partnership in science, technology and innovation (STI), the new project also offered a platform for stakeholders to share opinions and experience of Africa–Europe collaboration with the partnership's governance structure, the so-called EU–Africa high-level policy dialogue (HLPD) on STI, and of course with myriad national policymakers and programme owners.

Spanning a full decade (2008–2017), the CAAST-Net and CAAST-Net Plus projects became bywords for Africa–Europe collaboration, reinforcing the landscape, bringing together research and policy actors from the two regions to engage on topics of mutual interest and to conduct analytical work to advance the practice of cooperation. The projects have not been alone in this endeavour, however. CAAST-Net and CAAST-Net Plus joined its voice to that of a family of similar initiatives, many best known by their acronyms (PAERIP, PAEPARD, EUROAFRICA-ICT, RINEA to mention a few), over the past decade, which collectively have done much to build and reinforce our bi-regional S&T relationship in specific topics and as a cross-cutting domain.

This book has had a long gestation. Conceived in 2014 as a way of conveying the projects' learning in a more digestible and accessible format to a wider audience than its formal outputs, it also serendipitously fills the S&T gap in the existing body of literature on Africa–Europe relations. Writing not as academics but as practitioners, we have tried to bring together our collective practical experiences and analysis of cooperation in a way which we hope will provide a baseline for future assessment of our partnership, a guidance for international cooperation policy and programming and a sense of purpose to those working for a strong relationship that addresses shared societal challenges.

June 2017

Andrew Cherry
Eric Mwangi

Acknowledgements

The idea for a book about Africa–Europe research and innovation cooperation was first floated in 2014, in discussion between Gerard Ralphs and Andy Cherry. James Haselip and Isabella E. Wagner came on board soon after and took on central roles in managing the editorial process. However, the book reflects the rich experience and insights of all the members of the CAAST-Net and CAAST-Net Plus teams as well as the projects' close collaborators. As such, the editors recognise that the production of this book would not have been possible without the contributions of a wide range of individuals and their organisations. We would like to extend our specific thanks to the following:

The chapter authors and the authors of the outcome testimonials profiled through this book, for their commitment to the project, and for their patience from the early review stages through to the finalisation of the manuscript.

Anaïs Angelo joined the project team close to the book's finalisation as an editor. Her assiduous attention to detail and commitment to editorial excellence have been incredible assets, for which we are truly grateful.

The team from Palgrave Macmillan were a pleasure to work with. Sarah Roughley patiently supported and accompanied us right from the beginning, and Samantha Snedden, Oliver Foster and ArunPrakash Ramasamy assisted us very capably with the completion of the manuscript.

Financial support for the editing, layout and publishing of this book was kindly provided by the European Commission (EC), through the CAAST-Net Plus project (grant agreement 311806). The support of the EC has also helped to ensure that the book will remain an open access resource for readers across the globe.

Lastly, we are grateful to all of the partner organisations, partner representatives and stakeholders who have been part of the CAAST-Net and CAAST-Net Plus family over the past decade, for their contributions to strengthening Africa–Europe cooperation in research and innovation.

Andrew Cherry
James Haselip
Gerard Ralphs
Isabella E. Wagner

Contents

List of Contributors

Jean Albergel has been the director of France's CNRS-IRD joint office in South Africa. He has worked extensively as a scientist within the EU's Framework Programmes and coordinated the HYDROMED (FP4) and SOWAMED (FP6) projects. Within the context of EU research cooperation and support action projects, Albergel has been a work package leader in two Africa–Europe INCO-NET projects, CAAST-Net and CAAST-Net Plus, and also coordinated of the ERA-NET project ERAfrica.

Arlène Alpha has been a research fellow at the French Agricultural Research Centre for International Cooperation (CIRAD). A specialist in food and nutrition security policies in West Africa, her focus is on Burkina Faso. Alpha is also in charge of a research platform on food security in West Africa, which aims at developing research collaboration between CIRAD and West African public research institutions.

Ismail Barugahara heads up the Science, Technology and Policy Coordination Division at the Uganda National Council for Science and Technology. As one of the long-standing UNCST representatives within the CAAST-Net and CAAST-Net Plus projects, Barugahara has co-authored CAAST-Net research on the institutional landscape for Africa and Europe S&T cooperation, and on linkages between S&T and development cooperation.

Lauranne Botti is Manager of the Research Fairness Initiative at the Council on Health Research for Development (COHRED) and based in Geneva, Switzerland. She has worked as a COHRED representative within the CAAST-Net Plus project to advance the RFI as a sustainable platform for European and sub-Saharan African cooperation in research, development and innovation for health.

Alexandre Caron is a researcher at the French Agricultural Research Centre for International Cooperation (CIRAD), where he conducts research on disease ecology at wild/domestic interfaces in Transfrontier Conservation Areas in southern Africa. He is currently hosted by the Faculdade de Veterinaria at the Universidade Edouardo Mondlane in Mozambique, involved in the coordination of the Research Platform RP-PCP: "Production and Conservation in Partnership".

Andrea Cefis is Project Manager for Improving Food Safety in Benin, working closely with the Belgian Development Agency. He is a specialist in food safety and food security in tropical countries and has worked in several developing country projects to improve agro value chains.

Andrew Cherry has coordinated both the CAAST-Net and CAAST-Net Plus projects since 2008. He is an entomopathologist and trained at Imperial College London. He is presently based at the Association of Commonwealth Universities (ACU), where he was assigned initially to direct the Africa Unit and latterly to manage the ACU's projects on fostering international collaboration in scientific and technical research and innovation.

Nouhou Diaby is a researcher in the Institut Fondamental d'Afrique Noire at Université Cheikh Anta Diop de Dakar, Senegal. Diaby also serves as a technical advisor to the Ministry of Higher Education and Research. Since 2013, Diaby has been the focal point in Senegal for UNESCO's Global Observatory of Science, Technology and Innovation Policy Instruments (GO-SPIN).

George Owusu Essegbey is the director of the Science and Technology Policy Research Institute of the Council for Scientific and Industrial Research in Ghana. He has conducted extensive research on STI development, especially in agriculture and industry. His key thematic areas of research are micro and small enterprises, innovation studies, STI policy and climate change. He has also executed assignments for agencies of the UN system, including the World Bank, UNESCO, WIPO and UNEP.

Judith-Ann Francis is a senior programme coordinator, Science and Technology Policy, at the Technical Centre for Agricultural and Rural

Cooperation ACP–EU (CTA), The Netherlands. She is also the executive secretary of the European Forum on Agricultural Research for Development and represents CTA in two FP7 projects: CAAST-Net Plus and PACENet+. Both projects are bi-regional STI collaborations for addressing the global challenges on food and nutrition security, climate change and health.

Jochen Froebrich leads the Green Economic Growth Programme at Alterra Wageningen UR, The Netherlands. A specialist in water stress management in arid and semi-arid regions, his current focus is on transdisciplinary approaches for fostering agribusiness innovation, circular and green economy, as well as the related use of Public Private Partnership concepts in the Netherlands, Europe, India and Africa.

James Haselip is a senior researcher at the UNEP DTU Partnership and is based in Copenhagen. He works within the Cleaner Energy Development group, where his work focuses on the design and implementation of enabling frameworks for climate change mitigation technologies, using multi-criteria, economic baselines and outcome mapping methodologies. Since 2013, he has been involved in the CAAST-Net Plus project, contributing to its work on climate change.

Mike Hughes is an advisor for Science, Technology, Research and Innovation within the Ministry of Education, Rwanda. His key role is to develop Rwanda's national policy and strategy for the development of STI for poverty alleviation and economic growth. Hughes has also served in the Office of the President of Rwanda.

Carel IJsselmuiden a physician, epidemiologist, public health practitioner, academic and social entrepreneur, has been the executive director of the Council on Health Research for Development (COHRED). IJsselmuiden also teaches and undertakes research as an adjunct professor in the School for Applied Human Sciences at the University of KwaZulu-Natal in South Africa. He was formerly head of department of the Department of Community Health at the University of Pretoria, South Africa, and is a founding director of its School of Health Systems and Public Health.

Erika Kraemer-Mbula is an associate professor at the University of Johannesburg, South Africa, and Researcher at the DST-NRF Centre of Excellence in Scientometrics and Science, Technology and Innovation Policy. Her research interests concern the various routes involved in creating technological competencies in Africa. She recently co-authored *The Informal Economy in Developing Nations: Hidden Engine of Innovation?*

Katharina Kuss is an advisor on International Cooperation, Research, Health and Gender at the Spanish Foundation for International Cooperation, Health and Social Affairs. She has worked as an external evaluator for the European Commission, contributed to the development of several EU projects and serves on an international advisory board on women's health. She has coordinated, managed and participated in several projects funded by Directorate-General Research and DG Justice, including CPN-YAS-PRD, PRD College, Health NCP Net, CHANGE and CAAST-Net Plus.

Jacques Lançon is a senior advisor for African institutions and policies (Direction for Research and Strategy) at the French Agricultural Research Centre for International Development (Cirad). As a researcher, he has been responsible for projects and programmes on participatory plant breeding, multi-actor platforms and seed exchanges with various crops. He has also coordinated multidisciplinary and conceptual research on how to design new cropping systems, as well as instructed loans for the Development Bank of Vanuatu.

Toto Matshediso is Deputy Director for Strategic Partnerships at the Department of Science and Technology in South Africa. In this capacity, he has promoted and supported participation of South African National System of Innovation stakeholders in EU's programme of strategic importance to South Africa, including Horizon 2020, EUREKA, EDCTP, COST, Erasmus Mundus and the Africa, Caribbean and Pacific Science and Technology programme.

Priscilla Mugabe is an associate professor in the Department of Animal Science, Faculty of Agriculture, University of Zimbabwe and is an alternate coordinator for Research Platform Production & Conservation (RP-PCP).

Eric Mwangi is the deputy director of the Department of Research Management at Kenya's Ministry of Education, Science and Technology, and has been the Africa Region Coordinator for the CAAST-Net and CAAST-Net Plus projects. Through CAAST-Net and CAAST-Net Plus, Mwangi has fostered MOEST and Kenyan participation in FP7 projects, including ICT, research infrastructures and space science.

Emeka Orji is the deputy director of the National Office for Technology Acquisition and Promotion, Nigeria. He is a scientist and technology management professional, with extensive experience in technological innovation, S&T policy dialogues and strategies. He is also a monitoring and evaluation expert within the development of the small- and medium-sized enterprises sub-sector.

Melissa Plath is Head of Projects at the Finnish University Partnership for International Development, a network of Finnish universities supporting the attainment of strategic global responsibility objectives within the Finnish higher education sector. She is responsible for international cooperation and science policy within the network and manages UniPID's externally funded projects.

Erick Rajaonary a chartered accountant, is the president of the Malagasy Entrepreneurial Association and CEO of GUANOMAD. In 2013, Guanomad won the Outstanding "Small and Growing Business in Africa" within the African Leadership for Entrepreneurship in Mauritius.

Gerard Ralphs joined the Centre for Science, Technology and Innovation Indicators at the Human Science Research Council, South Africa, in 2017 as a programme manager and policy analyst. Prior to this he was the manager of partnerships and projects at Research Africa, also in South Africa, where he worked on the European Union-funded CAAST-Net, RIMI4AC and CAAST-Net Plus projects. In 2011–2012, he was based in Ottawa as a research awardee in the Donor Partnerships Division at Canada's International Development Research Centre (IDRC).

Jean-Michel Sers is European Affairs Coordinator for CIRAD. Prior to joining CIRAD, he was a policy officer at the European Commission's Directorate-General for Research and Innovation, and in charge of research and innovation cooperation between European and South Asian countries. In this role, his tasks included conduct and development of research and innovation policy dialogue with majors Asian country partners of EU, including the preparation of policy and position papers.

Mamohloding Tlhagale is the former head of International Cooperation and Partnerships at South Africa's Water Research Commission. She was formerly Director: Strategic Partnerships at the Department of Science and Technology, where she promoted strategic international partnerships to leverage international resources to strengthen South Africa's science and technology capacities.

Daan du Toit is Deputy Director General, International Cooperation and Resources, at South Africa's Department of Science and Technology. He has represented South Africa in various multilateral forums dedicated to international S&T cooperation, is a member of the Square Kilometre Array Organisation's Strategy and Business Development Committee and currently represents South Africa on the Joint Expert Group of the Africa–EU Science, Information Society and Space Partnership.

Arne Tostensen is a researcher at the Chr. Michelsen Institute, Norway, and is also seconded part-time to the Research Council of Norway. His research interests include human rights, development assistance, governance and democratisation, social security, poverty and institutional analysis, research policy and the research–policy nexus.

Constantine Vaitsas is the deputy head of the International Cooperation Unit at Forth/Praxi Network. He also worked for Rolls-Royce Plc and BTG Plc before focusing on International Cooperation and Innovation Management in Greece with exposure in a large number of countries and various types of stakeholders ranging from policymakers to SMEs and industry.

Johan Viljoen is the project officer for CAAST-Net Plus at the Institut de Recherche pour le Développement (IRD). He was also IRD's project manager for the FP7 ERAfrica project. Viljoen has also served as a project officer for researcher development and international relations at the South African National Research Foundation.

Isabella E. Wagner is a researcher and project manager at the Centre for Social Innovation (ZSI) in Austria. As a researcher, she evaluates science and technology policy programmes for ZSI. As a project manager, she is responsible for the planning and implementation of multimedia communication and dissemination strategies mainly for EU-funded cooperation and support actions in international science relations.

List of Abbreviations

ACP	African, Caribbean and Pacific Group of States
AfriAlliance	Africa–EU Innovation Alliance for Water and Climate
AMCOST	African Ministerial Council on Science and Technology
AMMA	African Monsoon Multidisciplinary Analysis
AR4D	Agriculture Research for Development
ASARECA	Association for Strengthening Agricultural Research in Eastern and Central Africa
AU	African Union
AUC	African Union Commission
AURG	African Union Research Grant
BCC	Business Cooperation Centre
CAAST-Net Plus	Advancing Sub-Saharan Africa–EU Research and Innovation Cooperation for Global Challenges
CAAST-Net	Network for the Coordination and Advancement of Sub-Saharan Africa–EU Science and Technology Cooperation
CCARDESA	Centre for Coordination of Agricultural Research and Development for Southern Africa
CIOMS	Council for International Organisation of Medical Sciences
CIRAD	Centre de Coopération Internationale en Recherche Agronomique pour le Développement
CNRS	Centre National de la Recherche Scientifique
COHRED	Council on Health Research for Development

CORAF	Conseil Ouest et Centre Africain pour la Recherche et le Développement Agricoles
CORDIS	Community Research and Development Information Service
COST	Cooperation in Science and Technology
CPA	Consolidated Plan of Action
DCI	Development Cooperation Instrument
DG DEVCO	Directorate-General for International Cooperation and Development, European Commission
DG RTD	Directorate General for Research and Innovation, European Commission
EC	European Commission
ECD	European Consensus on Development
EDCTP	European and Developing Countries Clinical Trials Partnership
EDF	European Development Fund
EEN	Enterprise Europe Network
ENDORSE	Energy Downstream Services
EU	European Union
FACCE-JPI	Joint Programming Initiative on Agriculture, Food Security and Climate Change
FAFS	Framework for African Food Security
FAO	Food and Agriculture Organization
FARA	Forum for Agricultural Research in Africa
FDI	Foreign Direct Investment
FNS	Food and Nutrition Security
FNSSA	Food and Nutritional Security and Sustainable Agriculture
FOCAC	Forum on China–Africa Cooperation
FP	Framework Programmes
GDP	Gross Domestic Product
GFATM	Global Fund to Fight AIDS, Tuberculosis and Malaria
GNI	Gross National Income
HACCP	Hazard Analysis Critical Control Point
HLPD	High Level Policy Dialogue
Horizon 2020	Horizon 2020, 8th Framework Programme for Research and Innovation
IMI	Innovative Medicines Initiative
INSERM	Institut National de la Santé et de la Recherche Médicale

IRD	Institut de Recherche pour le Développement
JAES	Joint Africa–EU Strategy
KEMRI	Kenya Medical Research Institute
KNUST	Kwame Nkrumah University of Science and Technology
KPFE	Swiss Commission for Research Partnerships with Developing Countries
LCSSA	Laboratory for Control of Sanitary Food Safety
MDG	Millennium Development Goal
MINWARE	Mine Water as a Resource
MoU	Memorandum of Understanding
NCD	Non-Communicable Disease
NCP	National Contact Point
ND	Neglected Disease
NEC	National Ethics Committee
NEPAD	New Partnership for Africa's Development
NRM	Natural Resource Management
NUS	Neglected or Underused Species
OECD	Organisation for Economic Co-Cooperation and Development
PACTR	Pan-African Clinical Trials Registry
PAEPARD	The Platform for African European Partnership on Agricultural Research for Development
PRD	Poverty-Related Disease
QWeCI	Quantifying Weather and Climate Impacts on Health in Developing Countries
R&I	Research and Innovation
REC	Regional Economic Community
RFI	Research Fairness Initiative
RP-PCP	Research Platform - Production and Conservation in Partnership
RROs	RFI Reporting Organisations
S&T	Science and Technology
S3A	Science Agenda for Agriculture in Africa
SDG	Substainable Delopment Goal
SME	Small and Medium-Sized Enterprise
SSA	Sub-Saharan Africa
STI	Science, Technology and Innovation

STISA	Science, Technology and Innovation Strategy for Africa
TB	Tuberculosis
UHC	Universal Health Coverage
UN	United Nations
UNESCO	United Nations Educational, Scientific and Cultural Organization
VicInAqua	Integrated Aquaculture Based on Sustainable Water Recirculating System for the Victoria Lake Basin
WABEF	Western Africa Biowastes for Energy and Fertiliser
WASH	Water, Sanitation and Hygiene
WECARD	West and Central African Council for Agricultural Research and Development
WHATER	Water Harvesting Technologies Revisited
WHO	World Health Organization

List of Figures

Introduction

Scientific and technological relationships between Africa and Europe have a long, dynamic and, sometimes, chequered history, which mirror an intricate array of national, regional and international interests and strategies. The nature of these relationships has invariably evolved over time. During the colonisation of Africa, European colonial powers not only battled for territory but also designed scientific and other policies so as to extract natural resources, establish new disciplines and generate botanical and often highly controversial human collections (e.g. see Dubow 1995, 2006; Shepherd 2003; Crais and Scully 2009). Shortly after the first wave of African independences in the 1960s, international aid for science was mostly directed towards establishing the first African universities, though the sector later fell into decline because of structural adjustment policies (Heidhues and Obare 2011). Today, in 2017, the dominant discourse is one of equitable partnerships between African and European nations, emphasising common interests, mutual benefits and global challenges (JAES 2007).

Over the past decade, Africa–Europe strategic partnerships have fuelled an expanding body of academic literature, which has attempted not only to assess the changing nature of such partnerships but also to foster the space for critical and creative reflection on the opportunities they offer to boost regional and global development. Adekeye Adebajo and Kaye Whiteman's recently edited volume *The EU and Africa: From Eurafrique to Afri-Europa* (Wits University Press 2013), for example, provides an extensive introduction on the historical, sectoral and geographical development of Africa–Europe cooperation. Returning to the colonial concept of "Eurafrique", they question whether Euro-African partnerships have ever been able to escape their imperial origins.

The complex relationships between a colonial past and the innovative potential of Euro-African cooperation in the twenty-first century have been increasingly scrutinised by scholars. Lukas Neubauer's *The EU-Africa Relationship: Development Strategies and Policies of the EU for Africa* (GRN Verlag Publishers 2010) has evaluated the legal foundations and principles that sustained the European Union's (EU) cooperation strategies in Africa since its beginnings in the 1950s. Jack Mangala's *Africa and the European Union: A Strategic Partnership* (Palgrave Macmillan 2012) assesses the successes and limitations of the Joint Africa–EU Strategy (JAES) adopted in 2007. Sectoral cooperation, too, has been analysed: Toni Haastrup's *Charting Transformation Through Security: Contemporary EU–Africa Relations* (Palgrave Macmillan 2013) has dealt with security cooperation, while Gerrit Faber and Jan Orbie's *Beyond Market Access for Economic Development: EU–Africa Relations in Transition* (Routledge 2014) has aimed to uncover the so far hidden faces of the Economic Partnership Agreements meant to stimulate trade between African countries and Europe.

Within this relative abundance of literature on bi-regional relations, research and innovation (R&I) cooperation between Africa and Europe has remained strikingly absent. Such a lack is perhaps surprising at a time when science, technological discoveries and the private sector are playing an increasingly pivotal role in shaping development policies, and when science, technology and innovation (STI) partnerships are becoming a priority area within many national and global development strategies. Therefore, an assessment of the scope of R&I cooperation, its structural and sectoral developments, the types of partners it involves (or excludes) and, perhaps more importantly, its potential to address the most urgent global issues is needed. It is precisely this gap that this volume addresses.

Scope and Audience

This book touches on several dimensions, admittedly with greater and lesser degrees of emphasis. It covers both Europe and Africa, but touches only the surface of the multiple partnerships that link both continents. It attempts to show the evolution of multilateral relations in the fields of scientific and technological research and cooperation, though glides over the issue of bilateral relations. It presents some of the leading multilateral

STI projects, their achievements and the persistent or future challenges they still face. STI cooperation between Africa and Europe defies boundaries, be they geographical or scientific, and, as such, is a vast topic that would deserve many more volumes.

Given the historical, geographical and strategic complexity of the Africa–Europe cooperation landscape, defining the scope of STI cooperation presented a challenging editorial task. Indeed, writing an exhaustive critical assessment of the topic seemed too big and too complex an enterprise. At the same time, the authors' individual professional and academic turfs risked being too narrow to be of wide interest or to faithfully capture the bi-regional nature of our subject. Recognising these limitations, we have worked to position ourselves in a "middle ground" to ensure this book convincingly addresses a much wider audience while reflecting the authors' expertise.

Turning more closely to the issue of scope, we embrace a broad definition of cooperation, referring to the pursuit of goals of common interest, following strategies defined and agreed on equal terms. In this context, the issue of funding is a recurrent element and therefore central to any discussion of Africa–Europe cooperation in R&I. Whether funded by international publicly owned research funding programmes or by national or commercial funding programmes, partnerships encourage broad multilateral participation across the two regions and involve researchers from multiple countries. Significant funding programmes that fit this categorisation include the EU's successive Framework Programmes (FP) for R&I, the European Development Fund (EDF) and the EU's Development Cooperation Instrument (DCI) (as in the case of South Africa–EU relations).

Multilateral cooperation does not of course exist in isolation and represents but a small portion of a large pie. Bilateral research, research funded by charities and philanthropic organisations, trusts and development banks, all together far exceeds the volume of multilateral research cooperation. While this book focuses mainly on multilateral cooperation supported by international programmes, we also draw on the experience of Africa–Europe cooperation funded in other contexts.

Thematically, as our subtitle suggests, this book deals with cooperation in applied research that seeks solutions for common, societal challenges and that fosters widespread societal benefit. It particularly emphasises the common narrative to Europe–Africa research cooperation in the twenty-first century, that of achieving mutual benefit through equitable partnerships. Such principles lie at the core of the JAES adopted in 2007, which has formalised an STI partnership between Africa and Europe. This book aspires to evaluate the purpose and future role of the new partnership and its relationship to the rest of the cooperation landscape. To be clear, while basic research is not specifically excluded from our conception of R&I in this book, our interest lies in the exploitation of new knowledge and the generation of technological innovations that emerge from applied science for social well-being and broad economic development.

Equally, we are interested in the nature and underlying process of Africa–Europe cooperation and in the essence of its bi-regional, multilateral partnerships. More specifically, we include considerations such as the conditions under which cooperation takes place—the framework conditions, the barriers that may hinder improved cooperation and the policy and programming responses that could enhance cooperation. Cognisant of the already large and growing global scholarly community that conducts large-scale STI surveys, such as research and development (R&D) surveys or business innovation surveys, a disclaimer is in order: we have not used repositories of STI statistics to develop the analysis, though we recognise that these data could provide an interesting archive for future research about the inputs into and outputs of the relationship. Instead, we have chosen to work with cooperation practitioners at the front line of cooperation efforts, drawing them in as chapter authors to reflect on their practice and to share their learning from the field. To the extent, then, that this book has been conceptualised as a practical resource, informed by a diversity of thinkers and "practitioners", rather than an advanced theoretical or empirical discussion, we hope that its analysis and content will be of interest to a diversity of readers and will inspire further research, critique and engagement.

Within this multi-layered set of concerns and constraints, many issues and questions are raised and, as much as possible, addressed throughout this book: Given the historical and political background of Africa–Europe cooperation, how does research cooperation support diplomacy in general, and scientific diplomacy in particular? Where do scientific relations fit into the bigger picture of Africa–Europe political relations? Furthermore, in what ways might the impact of technological innovation on scientific cooperation be more critically assessed? In an age of mass data flows, when international research is rapidly changing and researchers become increasingly mobile and have multiple affiliations, one is bound to ask whether concepts of nationality and regionalism retain their relevance and how these changes affect Africa–Europe cooperation: do they exist, in any meaningful way, beyond the concept, and is the Africa–Europe dichotomy losing its relevance on the global stage? How can publicly funded multilateral cooperation and commercially oriented private sector research find a common ground? Finally, does the rise of new scientific powers, the ascent of new strategic partnerships and even the emergence of the technologically literate classes announce a breakdown of disciplinary boundaries for bi-regional cooperation?

In sum, this book tackles four main challenges. First, it aspires to provide an accessible overview of the R&I policy landscape within which the Africa–Europe strategic partnership currently operates. Second, it undertakes to develop a critical analysis of the various networks and organisations that support, enable and enhance bi-regional STI cooperation. Third, it demonstrates the challenges of understanding the outcomes and impact of a number of cooperative STI initiatives. Fourth, it presents a series of clear lessons that can be taken forward to inform future Africa–Europe STI cooperation efforts.

A last point concerning issues of scope and definition is in order. We speak of "Africa–Europe" cooperation when referring to projects including various African and European countries *outside* of the formal institutional framework established by regional organisations, such as the EU and/or the African Union (e.g. in the case of the JAES). Furthermore, "Africa–Europe" serves as a more convenient appellation to ensure readers' attention does not get lost in the increasing complex constellation of partnerships. Similarly, STI and R&I should be read broadly and as synonyms, used interchangeably depending on the context and/or the specificities of the projects mentioned.

Outline of Chapters

The book is structured in three parts. The first part, "Policies, Politics and Programmes", establishes a description and critical analysis of the landscapes that have shaped and continue to influence the structure of Africa–Europe STI cooperation. The second part of the book, "Cooperation in Food Security, Climate Change and Health", gives centre stage to groups of thematic or sectoral specialists, who share their expertise and insider viewpoints as to how STI cooperation is responding to both policy priorities and ground-level challenges. The third part, "Futures of Africa–Europe Research and Innovation Cooperation", presents a set of forward-looking perspectives building from key themes explored throughout the book.

In Chap. 1 Daan du Toit and Andrew Cherry review the key political, economic and scientific drivers challenges for the Africa–Europe STI partnership. While emphasising that such cooperation is intrinsically influenced by a political, and even politicised context, the authors show that the JAES has proven to be a successful enterprise, despite the sceptics that continue to point to its institutional and structural limits. More importantly perhaps, the authors stress that its global nature has great potential to further shape international cooperation.

In Chap. 2, Ismail Barugahara and Arne Tostensen provide a chronological overview of developments giving shape to Africa–Europe cooperation on STI, and propose practical ways to further improve and rebalance its underlying principles. STI issues appeared rather late in Africa–Europe partnerships, they argue, and this is a critical detail for the shapers of future cooperation. This chapter also reminds readers that bi-regional cooperation bears the scars of the prior colonisation of the African continent and the concomittant deep establishment of unequal and (geographically) unbalanced cooperation patterns. In this context, they suggest that the recent renewed interest on STI has emerged along with a renewal of institutional partners, such as the African Union, which is increasingly asserting itself as a key regional leader in STI cooperation.

Although structural imbalance remains a priority, Chap. 3 focuses on some of the achievements of Africa–Europe STI projects. Its authors, Erika Kraemer-Mbula, Constantine Vaitsas and George Essegbey, highlight some "success stories", focusing specifically on new water and sanitation technologies and green chemicals. Such a practical approach nonetheless shows the intrinsically unequal cooperation patterns among African countries, while the private sector remains significantly under-represented in the cooperation. Future cooperation, they argue, should therefore focus on how to better market and disseminate STI products and services.

Addressing the issue of food and nutrition security (FNS), the authors of Chap. 4 argue that Africa–Europe cooperation still faces significant challenges. The practical implementation of measures supporting innovative solutions for food security are still lagging behind, they assert. Also, equal partnerships still need to be further institutionalised in order to become fairer and more enabling of positive change. Since FNS issues are heightened by structural socio-economic, political and even environmental constraints, they require extensive networks of research and institutional collaboration. Despite several achievements, persistent asymmetries continue to burden the achievements of FNS cooperation. The authors point to the strategic role that African countries could play first in cooperating with less developed European countries and in promoting alternative concepts of nutrition and environmental development on the global stage.

In Chap. 5, James Haselip and Mike Hughes critically assess the landscape of Africa–Europe cooperation on the topic of climate change. The authors argue that the complexity of R&I challenges for this issue calls for subtler collaborative programming and more rigorous evaluation. Critically, they emphasise the need for greater harmonisation between scientific and political priorities on climate change, and also point out that project goals should be much more precisely defined so as to ensure that results can be measured and solutions progressively improved.

Chapter 6 investigates the strategic benefits of global health collaboration programmes for Africa–Europe cooperation. Lamenting the lack of alignment or harmonisation of research priorities and cooperation patterns, its authors show how recent positive research development on health issues in Africa can foster more constructive and more balanced research partnerships with European countries and institutions. In this vein, the authors lobby for greater support for the Research Fairness Initiative (RFI), that is, as a promising emerging global standard for fostering fair and sustainable research partnerships and as a tool to establish more inclusive and better institutionalised framework for Africa–Europe cooperation on health development and innovation.

In Chap. 7, Gerard Ralphs and Isabella E. Wagner reflect on the issue of the "health" of cooperative STI projects, by drawing out the distinction between project efficiency, on the one hand, and partnership effectiveness on the other. In a context where partnering can be challenging—for reasons related to history or power imbalances—the authors propose a set of concrete applications to address these challenges during the partnering process. In doing so, they argue that using evaluative concepts, such as "partnership learning", are needed to better situate partners within the increasingly complex political, economic and cultural settings of STI collaboration projects.

Taking a bird's eye view on all the chapters, the postscript offers a set of critical perspectives on the framework conditions that shape Africa–Europe STI cooperation today. Reasserting the necessity of cooperation between the two neighbouring continents, it argues for more radical and innovative cooperative mechanisms, more commercially oriented funding models and a stronger "outcome thinking" mindset so as to ensure the sustainability of collaboration that can not only foster greater synergies between countries, research institutions and/or the private sector, but also enhance the welfare of the society at large.

Stories from the Field

To begin to assess and communicate the impacts of Africa–Europe STI collaborations, we invited managers of various Africa–Europe projects in the areas of health, food security and climate change to describe from their perspectives the *outcomes* achieved (i.e. any observable and verifiable change resulting from a project's direct results or outputs). A key instrument for development work and programme management, such "outcome thinking" documents the overall role and importance of STI projects and enables project managers, financial backers and/or donor agencies to attribute their work to observable impacts. Well-documented and verifiable outcomes, however, are typically accompanied with a degree of uncertainty, which is referred to as the "attribution gap". Project managers should be conscious that there are always competing claims crediting project or programme with observable impacts. As a result, the evidence of specific outcomes bridging outputs and impacts is all the more necessary.

The eight "Outcome testimonials" published in this book also draw attention to lessons learned that could benefit similar, future projects. Three of the testimonials cover advances within the broader operational landscape of Africa–Europe research collaboration, while the remaining five focus on specific projects within the themes of food and nutrition, safety and climate change. These outcome testimonials span a number of African and European countries and include a wide range of public, private and non-governmental partners in their presentation of examples of how bi-regional cooperation can help address global challenges.

Association of Commonwealth Universities — Andrew Cherry
London, UK
UNEP DTU Partnership — James Haselip
Copenhagen, Denmark
Human Sciences Research Council — Gerard Ralphs
Pretoria, South Africa
Centre for Social Innovation — Isabella E. Wagner
Vienna, Austria

References

Adebajo, A., & Whiteman, K. (2013). *The EU and Africa: From Eurafrique to Afri-Europa*. London: C. Hurts & Co.

African Union & European Union. (2007). *The Africa-EU strategic partnership: A joint Africa-EU strategy*. Available from: http://www.africa-eupartnership.org/sites/default/files/documents/eas2007_joint_strategy_en.pdf. Accessed 9 Dec 2017.

Crais, C., & Scully, P. (2009). *Baartman and the Hottentot Venus: A ghost story and a biography*. Princeton/Oxford: Princeton University Press.

Dubow, S. (1995). *Scientific racism in modern South Africa*. Cambridge: Cambridge University Press.

Dubow, S. (2006). *A commonwealth of knowledge: Science, sensibility and white South Africa 1820–2000*. Oxford: Oxford University Press.

Faber, G., & Orbie, J. (Eds.). (2014). *Beyond market access for economic development: EU-Africa relations in transition*. London: Routledge.

Haastrup, T. (2013). *Charting transformation through security: Contemporary EU-Africa relations*. New York: Palgrave Macmillan.

Heidhues, F., & Obare, G. (2011). Lessons from structural adjustment programmes and their effects in Africa. *Quarterly Journal of International Agriculture, 50*(1), 55–64.

Mangala, J. (2012). *Africa and the European Union: A strategic partnership*. Basingstoke: Palgrave Macmillan.

Neubauer, L. (2010). *The EU-Africa relationship: Development strategies and policies of the EU for Africa*. Norderstedt: GRIN Verlag.

Shepherd, N. (2003). State of the discipline: Science, culture and identity in South African archaeology, 1870–2003. *Journal of Southern African Studies, 29*(4), 823–844.

PART I

Politics, Policies and Programmes

CHAPTER 1

The Politics and Drivers Underpinning Africa–Europe Research and Innovation Cooperation

Andrew Cherry and Daan du Toit

Abstract This chapter provides a critical reflection on the achievements, over the last ten years, of the Africa–Europe partnership in science, technology and innovation (STI), following the introduction of the Joint Africa–EU Strategy in 2007. Building on the CAAST-Net experience and knowledge, the authors assess the multiple drivers (be they political, economic, scientific or even diplomatic) that boosted bi-regional cooperation on STI. In emphasising the political interests and constraints that significantly affect such cooperation, the authors show the rich potential of STI as a unique set of tools to address increasingly internationalised issues on the global scene.

A. Cherry (✉)
Association of Commonwealth Universities, London, UK

D. du Toit
Department of Science and Technology, International Cooperation and Resources, Pretoria, South Africa

A. Cherry et al. (eds.), *Africa-Europe Research and Innovation Cooperation*, https://doi.org/10.1007/978-3-319-69929-5_1

Keywords Joint Africa–EU Strategy • Africa–EU Summits • Science, technology and innovation • STI for development • R4D • Bi-regional partnership • Institutional relationships • Co-ownership • Bi-regional cooperation • Science diplomacy • Political and economic impact • Networked science • Human capital development • Policy and programme coordination

Introduction

The convening in Abidjan during November 2017 of the fifth Africa–EU (European Union) Summit of Heads of State and Government provides an appropriate moment to reflect on the achievements of bi-regional cooperation between Africa and Europe in science, technology and innovation (STI) over the past decade. For the purpose of this chapter, bi-regional cooperation relates to political and operational partnerships in STI and allied domains pursued within the ambit of the Joint Africa–EU Strategy (JAES) (see African Union & European Union 2010)—a political framework adopted at the second Africa–EU Summit held in 2007 in Lisbon. The chapter's focus is not therefore on the broad, diverse and long-term landscape of scientific partnerships between the two continents, which, due to its complexity and scope, would be difficult to meaningfully assess, but is on a relatively recent and discrete component of this landscape borne of the JAES.

Over the same ten-year period, the CAAST-Net platform, formally launched at the beginning of 2008, has developed a valuable repository of knowledge and information on aspects of Africa–EU cooperation and on the Africa–EU bi-regional STI partnership (see https://CAAST-Net-plus.org/, 2017; Africa–EU Cooperation 2017). CAAST-Net is a valuable, perhaps unique, resource in understanding the achievements and the challenges experienced by the bi-regional partnership.

The 2017 Summit will seek renewed commitment to our STI partnership, building on these achievements and challenges. The timing is opportune to draw on CAAST-Net's accumulated resources to better understand the complicated political, economic and scientific context in which bi-regional cooperation is being promoted. Such an understanding will inform recommendations to continue to enhance our future cooperation. Thus, it is largely through the lens of the CAAST-Net experience that we reflect on the politics and drivers underpinning Africa–EU cooperation in STI.

Overview of the Chapter

Reflecting on the Africa–EU STI partnership first necessitates a better understanding of political context at the time of its gestation and emergence, and the drivers which motivated the efforts to promote an STI partnership. Analysing these conditions, which are not necessarily the same for Africa and Europe, offers a deeper understanding of the nature of the evolving relationship, its strengths and its future potential, on the mobilisation of resources, and of its scope for influence on other dimensions of the Africa–EU relationship and JAES.

The STI partnership between Africa and Europe cannot be considered in isolation of wider political relations between the two regions, especially given the dominant role of the institutions of the African Union (AU) and the EU (and, to some extent, their member states) in promoting this cooperation. With resources invested in and decisions pertaining to bi-regional cooperation taking place almost exclusively at governmental level, the context for cooperation is intrinsically political. At times, this overtly political context has led to some frustration, particularly, for example, among those in the scientific community, not accustomed to such a process.

After a brief review of this political context, we discuss a range of drivers and objectives that we consider to have played a part in motivating bi-regional cooperation, along with political and economic considerations, the factors inherent to collaboration to advance excellence in science, as well as aspects related to the institutional relationship between the AU and the EU. We then assess the degree to which actual cooperation activities undertaken over the past ten years (and their results) correspond to the political context and to the drivers that informed both regions' commitment to the STI partnership. In doing so, we also consider the evolution of the political context and the drivers for cooperation over the past decade. We conclude with a glance to the future and, without pre-empting what follows, offer initial policy-level recommendations on how bi-regional cooperation might be further enhanced.

Nevertheless, it should be noted that the seeds sown during the past decade to promote bi-regional cooperation may only bear fruit in the years ahead. The existence of the formal bi-regional partnership within a wider landscape can hinder the direct attribution of outcomes and impact to political and programming efforts at promoting cooperation. Attribution is further complicated by significant time lag between cause and effect.

Our Africa–EU STI partnership is unique and fast evolving. Its place within, and relationship to, the wider cooperation landscape is complex. Although the assessment of the political context and drivers of the partnership at this ten-year milestone is timely and necessary, it has the potential to be equally complex. We have kept our approach simple, avoiding the detailed analysis that would be pertinent yet beyond the remit of this book. It is our intention that this brief assessment, albeit a highly subjective exercise, provides relevant background to the chapters that follow and offers a practitioner's perspective to students of Africa–Europe relations, helping to inform opinions of the achievements and merits of the past decade of partnership, and in formulating appropriate recommendations to improve our future cooperation.

The Political Context for Bi-regional Cooperation

Towards Equal Partnerships

The political context for Africa–Europe relations in 2007 was one that saw the gathering momentum of significant change to the typical post-colonial relationship between Europe and its former colonies that prevailed during the second half of the twentieth century. These years were marked by the Lomé Convention and subsequently by the dispensations of the Cotonou Partnership Agreement, with their primary focus on European development aid to Sub-Saharan Africa and preferential access to European markets for developing countries.

In 2007 discussions on the new Economic Partnership Agreements (EPA) between Europe and different African regions were in full swing, preparing the way for a relationship that would see a greater focus on reciprocity in African and European commitments, for example with regard to trade, and an emphasis on values such as co-ownership and co-responsibility.

In 2007, Africa was represented by a still relatively new AU, established in 2001, with a comprehensive focus on continental cooperation and integration. The development of a cross-sectoral inter-institutional AU–EU partnership was shaped largely by the convening of the 2007 Lisbon Summit. The first Africa–EU Summit, held in Cairo in 2000, lacked the focus of the second summit in 2007 on the development of a comprehensive partnership. Whilst there had been contact between the EU and the AU's predecessor, the Organisation for African Unity, those contacts were largely set within the post-colonial context of the second half of the twentieth century. In 2007, the European Commission (EC) found in the African Union Commission (AUC) a counterpart with which to construct a new strategic partnership. Efforts to promote and cement a bi-regional STI partnership will have played a part in solidifying the emerging institutional relationship between the two Unions—and their two Commissions.

While the inter-institutional relationship provides an important context for the STI partnership's emergence, other international relations will also have had an influence. Not least are the bilateral relations between African and European member states, as well as the engagement of individual countries with the bi-regional partnership. For example, the 2007 Summit was convened under the Portuguese Presidency of the EU. Portugal, a country with historic links with Africa, prioritised the Africa–EU partnership on the political agenda for its Presidency. Relations between the EU and Africa's Regional Economic Communities (RECs), the Africa, Caribbean and Pacific (ACP)–EU partnership, and different frameworks for Euro-Mediterranean cooperation, have each had an influence on the shaping of the bi-regional partnership. Thus the availing of financial resources to support the JAES STI partnership can be dependent on decisions of the structures governing these relationships—consider, for instance, the governance of the ACP–EU relationship and the provision of science and technology (S&T) funds for Africa–EU cooperation under the European Development Fund (EDF).

Global Consensus on STI for Development

By agreeing in 2007 to include a specific focus on STI in their new partnership, African and European leaders were aligning themselves with an emerging consensus on science for development at that time. The 2002 World Summit on Sustainable Development had explicitly recognised in its Johannesburg Plan of Implementation that science should be an instrument of and not a reward for development. The first decade of the twenty-first century thus saw intense activity at the policy level, in international forums such as the G8, the OECD, UNESCO or the World Bank on how to best harness STI for development (see, e.g. Watkins and Ehst 2008; Juma 2005). The Carnegie Group of G8 science leaders, for example, in 2006 had a dedicated outreach meeting with African partners. In 2002, the ACP Group of States and the EU convened a dedicated forum on research for sustainable development to consider appropriate investments from the EDF to build STI capacity building in ACP countries, a theme which would subsequently receive regular consideration under the JAES. The emergence in 2006–2007 of Africa's Science and Technology Consolidated Plan of Action can also be seen, in the wider context, as another component of this global consensus, giving practical issue to Africa's high-level objective of building strong S&T constituencies for socio-economic transformation.

The Lisbon Summit, in adopting the JAES, structured Africa–EU cooperation in different partnerships, with STI being grouped together with information and communication technologies (or the information society) and space in the so-called Eighth Partnership. The policy context, which informed this design, was the strong development policy focus on the narrowing of the digital divide especially in the aftermath of the World Summit on the Information Society held in 2003 and 2005, and in which the EU had actively participated. Europe's role as an historic provider of space-based technologies and services to Africa, European efforts to provide Africa with information and data products from Earth observation platforms, and perhaps ambitions to safeguard and expand this role, further cemented the inclusion of space in this framework.

At the time of the Lisbon Summit, despite the close economic and development cooperation links between Africa and Europe, relations between the two regions continued to be marked by political disagreements, often significant. Against this backdrop, the good news story, which cooperation in STI represents, unscathed by political differences or sensitivities with regard to trade or other controversies, meant that science diplomacy had also become a popular currency for the strengthening of the overall Africa–EU partnership. Indeed, in years to come, STI successes, no matter how modest, were often put in the spotlight when the successes of the JAES were to be celebrated.

Not to be confused with the 2007 Lisbon Summit of African and European Heads of State and Government, 2007 also saw the agreement of the Treaty of Lisbon, amending the original constitutional basis of the EU. What marks the Treaty as particularly relevant to our discussion of the Africa–EU STI partnership is the explicit inclusion among the treaty's articles of the objective of strengthening the EU's scientific and technological bases towards a European Research Area.

The explicit recognition of the need for continued strengthening and integration of the scientific and technological base in Europe, albeit for economic and industrial competitiveness, provided a sound argument for the inclusion in the JAES of an equivalent commitment to S&T. What is deemed essential for Europe, and indeed for the world, must surely be equally essential for Africa and for the new, heightened Africa–Europe political relationship in the JAES.

Evolution of Africa–EU Cooperation in Research for Development

At the time of the launch of the JAES, the research for development community, despite the broad political support for this agenda, had just started cooperation under the recently launched EU's Seventh Framework Programme (FP7) for Research and Innovation. Earlier FP had with some success included dedicated activities to fund research cooperation orientated to development outcomes between European researchers and their developing country partners (especially in areas such as health, agriculture and the environment). The new FP7, however, did not include such a specific activity but instead mainstreamed cooperation with developing countries across all themes of the FP—with developing country participants still being eligible for EU funding. The jury was out on how successful this new approach would be.

The year 2007 thus marked the beginning of a new era of sorts for Africa–EU science cooperation, with a focus on what many saw as a maturing partnership mainstreamed within a broader landscape of international cooperation programmes in science. However, that policy intent and the rhetoric co-existed with a requirement, on the part of several African countries, for concerted international assistance to develop essential STI capacities such as human capital and research infrastructure. The STI capacity building objective was included in the JAES but in the years to come cooperation efforts under the STI partnership were sometimes hampered or confused because of a misalignment between the goals of advancing excellence in science through cooperation as equal partners on the one hand, and European assistance for African capacity building on the other. This manifested itself most clearly in challenges to afford development assistance to Africa through programmes focused on mutual benefit through research cooperation.

As evidenced by the inclusion of STI in a dedicated partnership with information society and space, the adoption of the JAES also symbolised the broadening of Africa–EU science cooperation to a bigger community and portfolio—beyond the traditional, although extensive and successful history of cooperation in agricultural development research, for example. Timid statements of intent were made with regard to cooperation in emerging and industrial technology areas, but more often than not still within the context of science for development, for example, in the use of nanotechnology for water purification. Perhaps most significantly, the EU declared its intents to develop a dedicated STI policy dialogue with Africa (as it had launched with other regions) and saw the then African Ministerial Council on Science and Technology (AMCOST) as a potential counterpart for this endeavour. The first CAAST-Net project was funded, through the EU's FP7, to prepare and support such a policy dialogue.

Key Drivers Informing Bi-regional Cooperation

The JAES policy commitments adopted by African and European leaders in Lisbon in 2007 were informed and underpinned by a set of shared drivers for common objectives. In this section we consider the key drivers and objectives for the STI partnership, their relative importance, as well as the differences on the African and European sides.

The Global Consensus on Science and Technology

We referred in the previous section to an emerging global consensus in the first decade of the twenty-first century on the role of STI in development, and to the explicit inclusion in the Treaty of Lisbon to an objective of building the EU's scientific and technological bases. In short, the widespread acceptance that capacity in scientific and technological research, and in innovation, offered a route to industrial competitiveness, to economic growth, to sustainable development and to poverty alleviation provided a robust and timely argument to the architects of JAES for a chapter on science for sustainable development that was hard to refute. Thus, despite a shortage of resources, S&T together with space, and information and communication technologies found their place in the JAES as instruments of sustainable development alongside traditional domains for international political relations such as democracy, governance, human rights, peace and security.

Science Diplomacy

Although not an explicit driver at the outset, over time both the African and European sides saw, in the inclusion of STI in the JAES framework, potential for science diplomacy: the STI partnership reinforcing the bi-regional relationship via influence in other policy spheres. This contribution would include safeguarding and expanding an historic trading partnership, although as the difficult EPA negotiations in years to come would show, the two sides would harbour different ambitions, with Africa, for example, seeking greater access to the European agricultural market, and with Europe seeking to expand its presence in the African services sector. The global security context and Africa's role as a partner for Europe in the space sector, for example, were also seen as potential beneficiary spheres, at least from the European side, from investment in bi-regional STI cooperation.

Given the strategic significance and continued dominance of development cooperation as the focus of Africa–EU relations, both the African and European sides also harboured ambitions for the bi-regional STI partnership to have an influence on this domain. From an African perspective, there was a strong demand for the partnership to contribute to STI capacity building on the continent. While shared by the European side, the perspective was perhaps nuanced by a desire to see a new dimension added to the historic Africa–EU development cooperation relationship, one that would ensure greater efficiency and deliver greater impact.

As part of the portfolio of science diplomacy objectives, it was also foreseen that the bi-regional STI partnership would contribute to strengthening the AU–EU institutional partnership. As with capacity building, the institutional partnership objective was shared by both sides, again perhaps from slightly different perspectives. From a political angle, the EC could leverage STI cooperation to develop a privileged relationship with the new AUC, whilst the partnership with the EU also provided the AU with opportunities to develop its influence within the continental STI landscape.

Networked Science Knows No Borders

The sharing of resources, experience and expertise, especially to address shared challenges, or advance frontier science projects, has historically been regarded as a major driver for international cooperation in STI. Africa and Europe joining forces and resources to harness science to address major societal challenges such as climate change, energy security or pandemic disease was, thus, also one of the major objectives for JAES' commitment to bi-regional STI cooperation.

Few countries invest in international cooperation in science as a purely altruistic endeavour. Parties typically have to leverage their respective comparative advantages to make them attractive as a partner to the other (e.g. niche expertise in key technology domains or access to unique geographic conditions or resources). It is doubtful whether such strategic considerations informed the development of the bi-regional partnership in any significant manner, other than that Africa was certainly keen to leverage Europe's strong STI capacities. The nature or complexity of the multilateral cooperation relations is such that they can present greater challenges to leveraging national benefits than do bilateral cooperation relations, and offers a perspective on the relative merits of investing in multi- and bilateral relations.

International cooperation also plays a crucial role in developing human capital for STI. Most countries invest heavily in researcher training and mobility programmes with an international dimension, to ensure their next generation of researchers are equipped with global networks and perspectives. From an African perspective, the bi-regional partnership had as an objective to ensure exactly such support for human capital development in Africa. Whether investment in the bi-regional partnership was a major driver for Europe's own human capital development objectives is doubtful, although the publicly stated European policy objective of promoting Europe as a preferred destination for global research talent also applied in Africa.

Enhancing Cooperation: Ensuring Greater Efficiency and Impact

Cooperation in STI between Africa and Europe did not start with the bi-regional partnership agreed in Lisbon. Neither did the partnership have ambition to encompass all aspects of cooperation—especially bilateral relations. It is widely understood that the scope and scale of STI cooperation between African and European institutions in a bilateral context are far more extensive than in a bi-regional context and there was a conscious effort not to duplicate that landscape.

The JAES however did set a major objective of ensuring better coordination and synergy between programmes implemented at the multilateral (AU–EU) level and bilateral initiatives between member states. Addressing the goal would help avoid duplication and ensure greater efficiency, impact and return on investment. The JAES also had the objective of providing greater strategic direction to funding instruments, thus, not only ensuring alignment between different funding instruments and cooperation opportunities available for Africa–EU cooperation but also providing strategic input into the formulation of new cooperation programmes. The STI partnership shared these objectives too, aiming to improve the efficiency of cooperation, for example, with regard to the mobilisation of resources and enhancing impact.

Alignment of Bi-regional Cooperation with Supposed Drivers

Ten years on from the 2007 Lisbon Summit, it is an appropriate moment to reflect on the achievements of the bi-regional partnership. It is relevant at this juncture to distinguish between the policy-level bi-regional partnership in STI governed by the High Level Policy Dialogue (HLPD), and operational thematic research and research for development projects implemented by African–EU partnerships between institutional actors and which are funded by associated programmes or aligned to the objectives of the policy-level partnership. Whilst other chapters in this book cover the outcomes of operational projects, our assessment here concentrates mainly, although not exclusively on the policy-level partnership. We consider the extent to which the partnership's broader achievements have responded to the drivers discussed above.

Any perceived mismatch, however, between original intention and actual achievement should not necessarily be cause for criticism. The efforts in the years preceding 2007 to promote and build a structured, formal bi-regional partnership in STI were, in many respects, pioneering, and the vision was simple and strong. That said, it is inevitable with the benefit of ten years' hindsight, the initial well-intentioned vision and assumptions of the partnership's protagonists may seem naïve or simplistic. The pragmatic agility to adapt to the rapidly changing environment that has been an important feature of the partnership's success thus far and will continue to be so beyond the 2017 Summit.

Political and Economic Impact

The STI partnership has enjoyed substantial acclaim. Summits, ministerial gatherings and other high-level events held in the context of our bi-regional relationship over the past ten years have celebrated the successes of the STI partnership. Public awareness and communication efforts associated with the JAES often put the STI partnership in the spotlight as a flagship of cooperation. Although this celebration takes place in the absence of independent critical analysis, the partnership has achieved and continues to achieve a political purpose, which suits both parties: The STI partnership itself is a tangible demonstration of good-willed collaboration between Africa and Europe, working together at an enhanced policy and programming level for the global good.

Elsewhere, in other policy and programming spheres, within or even beyond the partnership, there have been varying degrees of influence of the STI partnership: There is, for example, no discernible evidence of an influence of the STI partnership on EPA negotiations conducted over the past decade. On the other hand, the JAES STI partnership has been particularly successful in terms of political impact in the strengthening of inter-institutional cooperation between the AUC and the EC. The STI partnership's well-defined governance structure has regular formal meetings. Through these meetings the two services have developed a close and privileged partnership, marked, for example, by staff exchanges. The engagement of member state representatives in governance mechanisms, however, with the exception of that of a relatively small core group, has struggled to function optimally. On the other hand, some critics have argued that the level of EC support to the AUC has been so significant—many of the programmes implemented by the AUC's S&T department receive EU funding support—that it set the African agenda and potentially undermines the AUC's independence, while also confronting Africa governments with the reality of investing to support AU programmes.

The inclusion of STI as a dedicated focus area in the JAES also had some success in providing a new dimension to the Africa–EU development cooperation partnership. It informed the availing of resources under the Development Cooperation Instrument, to support a range of initiatives such as the African Union Research Grants (AURG) programme implemented by the AUC. An increased allocation of funds for S&T from the ninth to the tenth EDF is said to be a response to the inclusion of a science partnership in JAES, while there was no impact, however, on resources under the EDF, availed to national governments and the RECs. The JAES science partnership does appear to have sparked a renewed interest in STI in the EC's dedicated services for development cooperation, after years of a relative lack of interest.

Expanding the Knowledge Base

The ten-year course of the JAES science partnership has seen the accumulation of an expansive portfolio of associated Africa–EU cooperation initiatives around scientific and technological research and innovation (R&I), particularly with a development focus. There is moreover a general consensus, albeit among interested parties, of an expansion of dedicated Africa–EU STI cooperation relative to the pre-partnership period.

Whilst a core tenet of the bi-regional partnership is the pooling of resources and the sharing of experience and expertise, a shadow on this otherwise positive situation is that resourcing of the portfolio of initiatives thus far has been skewed, with most funding originating on the EU side. The coordinated call for Africa is one example of a large contribution to the portfolio supported by the FP7. A noteworthy exception to this observation is provided by the ERAfrica call for proposals, discussed elsewhere in this book, which was funded jointly by a group of African and European national research and development agencies through a common pooling mechanism.

Although the JAES foresaw a stimulation of interest in R&I partnerships, most of the actual cooperation involved research cooperation between higher education institutions and publically funded organisations, with limited private sector involvement. International innovation partnerships are, however, inherently more difficult to promote than those with a primary orientation towards research, so this was a shortcoming not unique to Africa–EU cooperation.

STI cooperation during JAES has served Africa's human capital development and other capacity building objectives well, with a range of student training and mobility programmes, launched under the Erasmus as well ACP instruments. The bi-regional partnership has also seen valued investment in developing Africa's high-speed research networking capacity, a critical research infrastructure requirement.

Aligning Efforts

Strong AUC and EC involvement in the cooperation has not yet attracted strong sustained reciprocal interest from either African or European member states to co-invest and coordinate programmes under the umbrella of the JAES, as was foreseen, although the STI partnership has provided inspiration and additional rationale for national programmes in support of cooperation between Africa and Europe. The multilateral ERAfrica funding programme mentioned above, as well as the more recent LEAP-Agri joint funding programme, also inspired by the JAES STI partnership, has seen African and European research funders joining forces to fund collaborative research partnerships.

On another positive note, efforts since the 2013 Africa–EU HLPD on STI to focus the STI partnership's attention on the area of food and nutrition security and sustainable agriculture (FNSSA) promises to deliver results. An FNSSA R&I partnership foresees a flexible multilateral funding arrangement supported by African and European programme owners. We already see the leveraging of dedicated resources under at least three aligned research programmes, including the EU's Horizon 2020 FP. Efforts are being made to encourage coordination between public and private actors in this R&I partnership, although at this early stage without tangible results.

Evolution of the Drivers for Bi-regional Cooperation

In Abidjan at the 2017 Summit of Heads of State and Government, Africa and Europe will recommit to the JAES and the bi-regional STI partnership it includes. It is opportune to ask if the drivers, which informed cooperation in 2007, still apply and how they may have evolved.

Profound economic, political, environmental and social changes in Africa and Europe, alongside the international agreements and frameworks responding to these changes, such as the United Nations 2030 Agenda, provide a rapidly evolving context for Africa–EU relations, for the JAES and for the cross-cutting STI partnership. Africa and Europe certainly have much to gain from increased political and economic ties. At the same time, however, the additional bi-regional cooperation opportunities afforded to Africa by the Tokyo International Conference for Africa's Development, or by the Forum on China–Africa Cooperation (FOCAC), adjust Africa's perspective on JAES. Indeed, China is a dominant trade and investment partner for Africa and a more substantial analysis might consider how the STI component of JAES compares to that of the FOCAC.

Since 2011 and with the increasing impact of EU's development policy, the so-called Agenda for Change (see European Commission 2017a), the EU's approach to development cooperation is also evolving. There will be different focus areas and eligibility criteria. If the ambition in 2007 was to move beyond a donor–recipient relationship, it will be an imperative in 2017 (see European Commission 2017b) . Beyond 2020, and the expiry of the Cotonou Partnership Agreement, the relationship between Europe and the ACP Group of States will change significantly (see European Commission 2016) while knowledge promotion through the digital economy, STI will be a specific objective.

It is not only the geopolitics that is changing: The entire global enterprise of science is undergoing rapid transformation, perhaps most notably in the context of open science and open innovation, with traditional ways of cooperation discarded for more integrated, dynamic approaches. Open access to scientific data and research results, along with increased investment in e-infrastructures, will enable "networked science", shaping future Africa–EU cooperation in STI. The new bi-regional partnership must respond appropriately or risk obsolescence. Indeed investing in constant exploration and rolling out of new collaborative mechanisms is perhaps one of the partnerships greatest opportunities to contribute to the vigour of the overall Africa–EU landscape.

It would also be interesting to see if the AU's Science, Technology and Innovation Strategy for Africa (STISA-2024) meaningfully influences the design of the future partnership and what role the various national and regional STI strategies that African parties are developing will play. Questions to consider also include the role of the African scientific diaspora in cooperation frameworks and the influence of a coordinated European response to cooperation with Africa—as, for example, debated in the EU's Strategic Forum for International Cooperation.

It is unlikely that economic drivers, including trade and investment objectives, will have a more significant bearing on the future STI partnership than it had in the past. Return on investment, especially for taxpayers in difficult times, will be questions leaders would have to answer. Dynamics such as attracting research and development orientated investment by multinational companies and ambitions to be competitive in technology intensive industries could introduce elements of competition to the "strategic partnership".

Conclusion

Despite these changing dynamics, or perhaps because of them, bi-regional STI cooperation will more than ever be strategically relevant and important to the wider political partnership between the two regions. The role of STI as a domain with cross-cutting influence should receive attention in Abidjan. Other factors to consider in determining new drivers of cooperation include the evolution of the institutional structures and priorities of both the AU and the EU. Thus, for instance, were an African Space Agency or an African Research Council to become AU priorities for bi-regional cooperation, the nature of the partnership could be dramatically altered. Furthermore, within an integrated STI response to global challenges, for example, promoted in multilateral forums, the value addition of bi-regional cooperation as opposed to more inclusive multilateral cooperation will receive scrutiny and become a driver for cooperation in its own right.

Despite the lofty goals and flowery language of the 2007 Summit outcomes and other strategy documents, it is essential to maintain a realistic perspective with regard to expectations. In a complicated institutional landscape, fraught with political sensitivities, capacity constraints and other challenges, the bi-regional STI partnership was never going to change the world. It never pretended to. As the chapters in this book show, in its first ten years the STI partnership, at both the policy and operational levels, has achieved important successes. These are worth celebration. Perhaps most significantly the investments over the past decade will continue to bear fruits in years to come as they have enabled a more robust and stronger STI partnership.

In a world where multilateralism and solidarity are precious commodities, Africa–Europe bi-regional cooperation continues to matter. STI, because of its cross-cutting impact and strategic significance, should play an ever more central role in this broad political relationship. Africa and Europe should aim to harness this potential, but a dedicated focus, with dedicated instruments to advance cooperation is required, as provided for the bi-regional partnership.

This book will conclude with a more comprehensive analysis on future developments. We hope that it shows that the decision in 2007 to invest in a dedicated bi-regional STI partnership was a correct one. The partnership has achieved much short-term tangible success at the project and policy partnership levels, as well as likely long-term though less tangible impact. However, an honest, perhaps even politically incorrect analysis, without favour or fear, to identify the shortcomings of the past decade as this book attempts, should play an important part in informing post-Abidjan plans—plans which should marry ambition with realism.

References

Africa-EU Cooperation. (2017). *Science, technology and innovation portal.* Available from: http://www.rinea.org/en/index.php. Accessed 27 June 2017.

African Union & European Union. (2010). *Joint Africa-EU Strategy: Action plan 2011–2013.* Available from: http://www.africa-eu-partnership.org/sites/default/files/documents/03-JAES_action_plan_en.pdf. Accessed 8 May 2017.

European Commission. (2016). *Joint communication to the European Parliament and the Council. A renewed partnership with the countries of Africa, the Caribbean and the Pacific.* Available from: https://ec.europa.eu/europeaid/sites/devco/files/joint-communication-renewed-partnership-acp-20161122_en.pdf. Accessed 25 June 2017.

European Commission. (2017a). *EU Communication on the agenda for change.* Available from: https://ec.europa.eu/europeaid/policies/european-development-policy/agenda-change_en. Accessed 27 June 2017.

European Commission. (2017b). *Joint communication to the European Parliament and the Council for a renewed impetus of the Africa-EU partnership.* Available from: https://ec.europa.eu/europeaid/sites/devco/files/joint-communication-renewed-partnership-acp-20161122_en.pdf. Accessed 25 June 2017.

Juma, C. (Ed.). (2005). *Going for growth: Science, technology and innovation in Africa.* London: The Smith Institute.

Watkins, A., & Ehst, M. (Eds.). (2008). *Science, technology and innovation: Capacity building for sustainable growth and poverty reduction.* Washington, DC: The International Bank for Reconstruction and Development/The World Bank.

CHAPTER 2

Policy Frameworks Supporting Africa–Europe STI Cooperation: Past Achievements and Future Responsibilities

Ismail Barugahara and Arne Tostensen

Abstract This chapter provides an overview of the chronological developments of the Africa–Europe cooperation on science, technology and innovation (STI). The authors first remind readers that African–European cooperation bears the mark of the prior colonisation of the African continent and the deep establishment of unequal and (geographically) unbalanced cooperation patterns. They then reflect on the renewed interest in STI in Africa–Europe cooperation. The chapter explains how more balanced cooperation structures and more efficient financing systems can be achieved and how science can become an integrated part of global development strategies.

Keywords Framework programme • Africa–Europe economic agreements • International cooperation strategy • Sustainable cooperation • Global development • STI cooperation

I. Barugahara (✉)
National Council for Science and Technology, Kampala, Uganda

A. Tostensen
Research Council of Norway and Chr. Michelsen Institute, Bergen, Norway

A. Cherry et al. (eds.), *Africa-Europe Research and Innovation Cooperation*, https://doi.org/10.1007/978-3-319-69929-5_2

Introduction

Science, technology and innovation (STI) cooperation between Europe and Africa has undergone significant change in the past decade. For example, by September 2013, towards the close of the EU's Seventh Framework Programme (FP7), there were 1315 participants from 45 African countries in 565 projects, with a total grant funding of 178 million euros from the European Commission (EC). About 82% of these projects received financing mainly through the FP7 "Cooperation Specific Programme" and were largely centred on health, agri-food and the environment.

The adoption in 2007 of the Joint Africa–EU Strategy (JAES) prepared the ground for partnerships based on egalitarian relationships and mutual benefits while acknowledging the critical role STI can play in boosting both human and socio-economic development (African Union & European Union 2007a). Nevertheless, cooperative efforts must be strengthened. More particularly, both European and African countries must continue to join forces to further align STI cooperation politics, harmonise their monitoring instruments and better coordinate their programmes.

This chapter first describes the prominent regional and bi-regional policy frameworks that underpin Africa–Europe cooperation on STI. Our chronology of Africa–Europe cooperation begins in 1957 (with the Treaty of Rome) and continues to more recent developments, with the adoption, in 2007, of the JAES and the establishment of a bi-regional partnership on food, nutrition security and sustainable agriculture (FNSSA) in 2016. It then focuses on two key issues that Africa–Europe STI cooperation is currently facing: first, the necessity to find new models for more equitable and sustainable funding of STI in Europe and Africa alike; and second, the need for continuous efforts to prioritise science and technology in the global development agenda (Agenda 2030). This chapter argues that if policymakers do not give these issues earnest consideration, cooperation risks not only forgoing the important gains it has made but also faces the risk of being weakened.

The Policy Frameworks

The Cotonou Agreement

In 1957, in the context of both African decolonisation and the early stages of building the modern-day European Union (EU), the Treaty of Rome made a provision for the creation of a European Development Fund (EDF) with a view to granting technical and financial assistance to African countries, some of which at that time were still colonies of European powers. The Cotonou Agreement paved the way for greater Africa–Europe cooperation in broad terms and, until today, serves as an umbrella under which many funding instruments fall. Preceded by the four successive Lomé Conventions (Lomé I–IV), the Cotonou Agreement was concluded in 2000 for a 20-year period as a partnership between the EU and 79 countries of the African (excluding South Africa), Caribbean and the Pacific (ACP) regions. Its fundamental principles include the equality of the partners, as well as inclusive participation, dialogue and regionalisation.

The Cotonou Agreement was revised in 2005 and 2010 and now rests on three pillars, namely, development cooperation, political cooperation, and economic and trade cooperation. The 2010 revision was particularly significant because it recognised climate change as a global challenge, committing the parties to include it in development cooperation and to support ACP efforts in mitigating and adapting to the effects of climate change. This revision was also instrumental in putting into practice the internationally agreed aid effectiveness principles laid down in the Paris Declaration, especially the principle of donor coordination.

Although the EDF falls outside the EU's Multiannual Financial Framework (MFF), it is the main source of funding under the Cotonou Agreement, and mainly directed to development cooperation with African governments, either through national programmes or through Africa's Regional Economic Communities (RECs). Financed by direct contributions from EU member states according to contribution shares (or "keys"), it is governed by its own rules. The current 11th EDF has 30.5 billion euros at its disposal for the period 2014–2020. The EDF operates thematic development cooperation programmes in areas such as health and the environment as well as programmes with a pan-African focus. While the EDF covers all ACP countries, the large number of African states forms a majority. A few national or regional activities funded by the EDF have a specific science and technology capacity building orientation.

The European Consensus on Development

The European Consensus on Development (ECD), adopted in 2006, formed the common basis of EU development policies and actions. The ECD was adopted in March 2005, aiming at eradicating poverty and promoting sustainable development. In-depth political dialogue was set out as an essential ingredient for good governance, the respect of human rights and the rule of law, the fight against corruption and spread of democratic rule. Furthermore, to address the problems of fragile states, the EU pledged to engage in institution building and to foster linkages between emergency aid, rehabilitation and long-term development.

The ECD reiterated its support for global initiatives such as the Millennium Development Goals (MDGs)—today replaced by the Sustainable Development Goals (SDGs)—covering key development issues such as poverty, health, food security, access to education, gender equality and environmental sustainability, and which emphasised the need of a global partnership for development. Just like the MDGs, the SDGs involve various sectoral and thematic domains of STI, whose advancement and application are implicated in their realisation. At the same time, it reaffirmed the need to coordinate with the Bretton Woods institutions' policies and programmes and to enhance cooperation with the United Nations (UN) system and other relevant institutions such as the Development Assistance Committee of the Organisation for Economic Co-operation and Development (OECD), especially the Good Practice Guidelines of the latter.

In terms of the volume of resources put at the disposal of development partners, the EU adopted a timetable for its member states to reach the 0.7% of gross national income (GNI) goal by 2015, with an intermediate collective target of 0.56% by 2010. If these targets had been achieved, it would have doubled EU development assistance to 66 billion euros by 2010. Although failing to achieve these targets, the EU has remained a major development partner for Africa. From 2007 until 2013, official development assistance in the form of national budget and project support disbursed to Africa by the EU and its member states was estimated at 144 billion euros, or about 20.6 billion euros on average per year (European Commission 2016c). The EU is committed not only to providing aid (with efficiency and quality) but also to entering into economic and trading partnerships with developing countries. It is worth mentioning that although the ECD did not prioritise STI cooperation, it has remained open, in principle, to such a collaboration.

Consolidated Plan of Action (2006–2014)

Africa's Science and Technology Consolidated Plan of Action (CPA) was adopted in 2005 to promote and use science and technology to foster socio-economic development and ensure Africa's integration into the world economy (NEPAD 2006). At the time of its formulation, the

African continent had made only modest progress in the area of STI, owing to numerous systemic constraints: weak inter-institutional linkages and collaboration; disjointed STI and development policies; weak policy implementation capacity; modest national allocations for science and technology (below 1% of gross domestic product (GDP)); as well as a limited capacity to translate research results into industrial products and services (Barugahara and Tostensen 2009a).

To address these challenges, the CPA established three priority areas: capacity building, knowledge production (as well as scientific research) and technological innovation. The following goals were set: (1) implementing the African Science, Technology and Innovation Indicators Initiative; (2) improving regional cooperation in STI (through capacity building, exchange of good practices, the formation of a common African framework for cooperation in science and technology, and a more active participation in EU facilities); (3) building a public understanding of STI; (4) building a common African strategy for biotechnology; (5) building science and technology policy capacity; (6) creating technology parks.

The CPA was implemented through networks of centres of excellence, which were dedicated to specific STI and capacity building programmes, while complementing other African Union (AU) and New Partnership for Africa's Development (NEPAD) specific programmes for agriculture, environment, health infrastructures, industrialisation and education. The AU Commission (AUC) provided overall political and policy leadership for the implementation of the CPA, while the NEPAD office for science and technology and the African Ministerial Council on Science and Technology (AMCOST) provided overall technical and intellectual leadership.

The CPA successfully managed to raise awareness among many African governments about the usefulness of STI in societal transformation and development processes. It also unified the so-far fragmented national systems of innovation (NSI) of the countries that had jointly formulated STI priorities. While the CPA was elaborated to address STI challenges in Africa, it was acknowledged, very early on, that a more thorough engagement with the EU would be advantageous to achieve its ambitious goals. Such engagement encompassed professional collaboration in research and innovation (R&I) activities, but also funding and capacity building in prioritised fields (Barugahara and Tostensen 2009a, p. 44).

Science, Technology and Innovation Strategy for Africa (STISA-2024) (2014–Present)

In 2014, the AU further committed itself to support STI policies and, in so doing, set up the Science, Technology and Innovation Strategy for Africa (STISA-2024), which was to replace the CPA. STISA-2024 was adopted during the 23rd Ordinary Session of the AU Executive Council during the AU Summit held in Malabo, Equatorial Guinea, and served as the continental framework for accelerating Africa's transition to an innovation-led, knowledge-based economy (EX.CL/839[XXV]). As an integral part of the Agenda 2063 set by the AUC, which recognised STI as a major driving force for socio-economic development, STISA-2024 was designed as a ten-year incremental strategy and emphasised the necessity to integrate STI into critical sectors such as agriculture, health, infrastructure development, mining, water, energy, and environment. Its six key priority areas included (1) eradication of hunger and achieving food security, (2) prevention and control of diseases, (3) communication, (4) protection of our space, (5) Living Together—Building the Society and (6) wealth creation. The meeting specifically called upon African states and the RECs to integrate STISA-2024 into their STI development agendas for implementation.

The EU's International Scientific Cooperation Strategy

International cooperation is expected to foster new knowledge production, increase scientific quality and improve the competiveness of R&I systems. At the same time, internationalisation boosts the productivity of investments in research and development by enabling companies to gain more knowledge from international markets, to participate in new value chains and reap greater benefits from growing markets outside the EU (European Commission 2016c).

Premised on the above expectations, in 2012, the EU adopted a strategy for international cooperation in R&I. The main objectives of the strategy are to strengthen the EU's R&I excellence, attractiveness and economic and industrial competitiveness, tackling global societal challenges, and supporting the EU's external policies. The research programmes carried out by the EU are open to participation by research institutions and researchers worldwide and cooperation is fostered through FPs for R&I (currently Horizon 2020). The EU is also developing targeted strategies with specific countries in order to achieve specific objectives: in 2015 the European Commission published 11 multiannual roadmaps for scientific cooperation with industrialised countries (Canada, South Korea, USA, Japan), emerging scientific powers (Brazil, Russia, India, China, South Africa) and the European Neighbourhood Policy countries in two groups (Eastern Partnership and Southern Mediterranean). Each roadmap presents the state of cooperation with the EU and defines thematic priorities for future cooperation in R&I (European Union 2015). Several other roadmaps with third countries, especially countries in Africa, have since been published.

As a result, R&I cooperation has been prioritised and intensified with the aim of making the EU a stronger global actor in solving global challenges in the areas of health, food, energy, water and climate change. Notably, the results of the EU's R&I efforts have contributed to the development and implementation of a number of international commitments such as the United Nations Framework Conditions on Climate Change, the Convention on Biological Diversity, the 2030 Agenda for Sustainable Development and various resolutions of the World Health Organization. More than 1000 publications from FP7 projects contributed to the fifth Assessment Report of the Intergovernmental Panel on Climate Change that provided the evidence base for negotiations at the UN Climate Change Conference held in Paris in 2015 (European Commission 2016c).

The Joint Africa–EU Strategy (2007–Present)

From a starting point of largely separate continental approaches to STI strategies, the JAES specified the terms of engagement between the two continents, aiming to strengthen political partnership and cooperation. The JAES' Eighth Partnership on Science, Information Society and Space (also discussed in Chap. 1) specifically recognised STI as necessary to fostering knowledge-based societies, competitive economies and sustainable development. The JAES policy framework came with a first action plan (2008–2010), later succeeded by a second one (2011–2013) reinforcing the commitments on STI, and reiterating the strategic importance of modern technologies to achieve the MDGs, and the SDGs subsequently, set up by the UN 2030 Agenda (African Union & European Union 2007b, 2010).

In 2014, the Fourth EU–Africa Summit reaffirmed that the JAES continues to frame continent-to-continent cooperation. On the same occasion, a roadmap for 2014–2017 was adopted for the implementation of the joint strategy, and redefined the priority areas as (1) peace and security; (2) democracy, good governance and human rights; (3) human development; (4) sustainable and inclusive development and growth and continental integration; (5) global and emerging issues. The third priority area concerned STI more specifically, while the roadmap made an unequivocal case for its role in shaping the relations between the two continents: "Investments in science, technology and innovation (STI) are vital to promote growth and employment, improve competitiveness and tackle pressing global challenges" (EU-Africa Summit 2014). The Summit also recognised the EU–Africa High Level Policy Dialogue (HLPD) on STI as an instrumental actor in the implementation of the STI part of the programme. Most recently, in April 2016, the College-to-College meeting between the European Commission (EC) and the AUC reaffirmed its commitment to continued collaboration to maximise the mutual benefits of STI towards addressing multiple challenges, including poverty (European Commission 2016a, b, c).

The inclusion in the JAES of STI cooperation reinforces the wide range of collaborative relationships that exist outside of the bi-regional policy framework. These include the relations between the EU and/or its member states with sub-continental groupings such as the RECs. Several European states have maintained long-standing relationships of significant magnitude with African counterparts, under the Commonwealth or similar groups. The rapidly changing global scientific, technological, socio-economic and political landscape has motivated current efforts towards the strengthening of the EU–Africa partnership.

All of these efforts have shaped the agenda for the Fifth Africa–EU Summit, which is scheduled for November 2017 with the aim to review and deepen the Africa–EU partnership. In the Joint Communication to the European Parliament and the Council for a renewed impetus of the Africa–EU Partnership, the EU proposes a revitalised framework for joint action. It sets out policy priorities and concrete initiatives for 2018–2020 and beyond, to be developed jointly with African partners, in response to Africa's Agenda 2063 and building on the Global Strategy for the EU's foreign policy (European Commission 2017).

The quest for a strengthened Africa–EU partnership is based on shared values and interests enshrined in the JAES. It is also based on a thriving long-term partnership where the EU remains Africa's significant development partner as illustrated by the following indicators: (1) the EU is collectively Africa's main foreign investor (32 billion euros of EU Foreign Direct Investments (FDI) flowed to Africa in 2015 (33% of total FDI flows to Africa); the EU accounted for 33.5% of Africa's imports and 42% of Africa's exports in 2016—the European Investment Bank (EIB) also provides over 2 billion euros of annual financing in Africa; (2) the EU is Africa's main trading partner, offering free access to the EU market for all products via Economic Partnership Agreements, the Free Trade Agreements and the Union's Generalised Scheme of Preferences; (3) the EU is the main source of remittances (21 billion euros of remittances from Europe to Africa in 2015 [36% of global flow to Africa]); (4) the EU is the first partner in development and humanitarian assistance (21 billion euros of collective Official Development Assistance (ODA) [EU and its member states]) to Africa in 2015 (50% of total ODA to Africa) (European Commission 2017).

Building on these concrete achievements, the EU suggested the JAES (2018–2020) should focus more strongly on sustainable and inclusive economic development in Africa by creating jobs and highlighting the opportunities it offers to Europe. The proposed flagship actions on R&I cooperation include: (1) launching a new EU–Africa R&I partnership on climate change and sustainable energy focusing on deployment as well as capacity building in renewable energy and energy efficiency and in climate services; (2) generating EU and African investments to support R&I in agriculture via the EU–Africa Research and Innovation Partnership on FNSSA as well as increase the uptake of new technologies by local communities for increased agricultural income and nutrition; (3) intensification of Africa–EU collaboration on research by (i) facilitating collaboration between researchers and innovators from Africa and Europe, including through increasing professional development opportunities for researchers through the Marie Sklodowska-Curie Actions and other types of Horizon 2020 projects, (ii) supporting research capacity building in Africa through programmes such as the AURG, and (iii) supporting an open digital research environment for universities and research organisations. These are plausible actions that when concretised and implemented will strengthen Africa–EU R&I cooperation and put both continents on a higher trajectory of development.

Three Key Policy Issues

Re-balancing Cooperation

There is robust evidence worldwide that greater use of technology is a major factor in enhancing productivity-driven growth and industrial competitiveness (Solow 1956; Temple 1999; Barro and Sala-i-Martin 2004; Romer 2007). What does this evidence mean for Africa–Europe STI cooperation, in the face of global challenges such as climate change, food security and human health? Apart from considering STI as an engine of growth, the EU has an interest in the creation of markets for its products in Africa, where competition from China, Japan and India is increasingly being felt. Economic growth and employment creation in Africa are also likely to ease the migration pressures EU countries are presently facing. In the event of higher growth, job creation and economic stability in Africa, as well as better adaptation to climate change, the anticipated need for recurring emergency relief operations will be reduced, as will be the threat of tropical diseases through the development of new medicines and vaccines, clinical trials and laboratory experiments. For the EU, these are among some clear and unequivocal reasons to cooperate.

Many African countries have recorded impressive economic growth rates in recent years, but this growth is likely to taper off unless buttressed by genuine economic transformation and technological advances (Booth and Therkildsen 2012). Cooperation between Africa and Europe can help both sides to achieve their objectives more effectively and efficiently than if they work in isolation from each other. However, more attention is required to address several systemic and structural challenges that tend to limit the continents' ability to effectively participate in R&I cooperation with the EU and the rest of the world (Barugahara and Tostensen 2009a). To ameliorate this situation, resources need to be leveraged in order to strengthen the capacity of African institutions and researchers to become effective and genuine partners with their EU counterparts in building robust STI systems across Africa, and in the EU. More effort is required to augment the existing Africa–EU collaborations, including CAAST-Net Plus which designed and implemented a series of research capacity building events to complement the work of related initiatives such as AfricaLics, whose secretariat is based at the African Centre for Technology Studies (ACTS) in Kenya, and the DfiD-funded Climate Impacts Research Capacity and Leadership Enhancement (CIRCLE) programme, implemented by the African Academy of Sciences. Policymakers therefore need to place greater emphasis on creating incentives and building adequate network infrastructure and human resource capacities that are required for effective participation in international cooperation activities.

Financing Cooperation

The appropriate modality of financing for various EU and African STI initiatives should involve a mix at multiple levels. The current practice with direct funding of national activities alongside funding through the RECs as well as through continental bodies may facilitate budgetary control and cost-effectiveness. Mechanisms should be put in place to ensure harmonisation of interventions at various levels for greater development impact and uptake of research findings with more efficient use of resources.

Specific project support measures for short-term research and/or development activities in the mutual EU–Africa STI priorities as given in the JAES should be emphasised. Longer-term and sustainable funding arrangements might be established through, for example, regional financial institutions such as the African Development Bank for regional and continent-wide STI initiatives, complemented by funding sources at the national level. Proposals towards that end have been mooted in several fora and need to be reinforced and made operational.

Bridging the Gap Between Science and Global Development

The world is facing formidable development challenges, above all reducing poverty and food insecurity. Meeting those challenges are the overarching objectives of development cooperation, and STI is a critical means to reach them. Although this book deals with bi-regional cooperation in STI, such cooperation is not an end in itself. It is indeed a means to an end. Hence, STI endeavours must be considered in conjunction with development efforts. So the uptake and application of research results is pivotal.

There is a plethora of policies and instruments related to STI, on the one hand, and development issues, on the other—at both EU and AU ends. Those policies and instruments need to be integrated into a coherent whole with a view to informing and underpinning development efforts with technologies stemming from research projects. This is not an easy task. Bridging the gap between STI and development efforts has been the subject of a long-standing and continuing discussion (e.g., Court et al. 2005; Leach et al. 2008).

A permanent forum for Africa–Europe dialogue was established in 2014 with the HLPD on STI and the adoption of a roadmap for cooperation. The regular College-to-College meetings serve the same purpose of ensuring political commitment and technical follow-up on the implementation of the strategy. These fora will provide the needed follow-up of the policy aspirations of the continental authorities and provide an arena for professional interaction among policymakers, technocrats and development practitioners across Africa and the EU as an essential first step towards bridging the gap between science and global development (Barugahara and Tostensen 2009b; Diyamett 2008).

The relative importance and priority (in terms of resource envelopes) accorded to research as distinct from development activities requires conceptual attention. While debate is rife between proponents of both aspects, it should be noted that the road from research to products on the market shelf is not as straightforward (or "linear") as it may seem in the case of development interventions. Taking this into consideration, a greater appreciation of the uncertainties of research endeavours is required when balancing the funding priorities between STI and development interventions. Sustainable and adequate funding mechanisms could be explored within the development cooperation instruments with assurance for higher returns from investments in R&I in the longer-term perspective.

The interface between STI and development efforts requires the building of operational models for bridging the existing gap. The models of collaboration between policymakers, scientists and practitioners with a view to achieving a greater uptake of research findings for development ends are essential components of development cooperation. This is only possible if the models are workable; that is, involving all relevant stakeholders operating on a common understanding and within a policy environment conducive to such collaboration. It would probably not be possible to arrive at a generic collaborative model that would fit all circumstances and sectors. Hence, most models, while replicable in some respects, will need to be "customised" to the specific conditions at hand, be they institutional or otherwise. ERAfrica is a new platform for research collaboration between Africa and Europe, co-funded by the EU and the collaborating partners, at a ratio of 20:80. The ERAfrica and the associated LEAP-Agri project model might herald the establishment of such innovative and viable models of development cooperation. Greater emphasis should be placed on the application aspects of Horizon 2020 projects in order to enhance the uptake of research outputs.

Conclusion

The socio-political, humanitarian, economic, and technological drivers of Africa–Europe cooperation are well articulated in the ECD (2006), the CPA (NEPAD 2006), the JAES (African Union & European Union 2007a) and the STISA-2024. Encompassing the aspirations of the two continents in development and STI, these policy documents constitute key elements of Africa–Europe cooperation. All of these frameworks note the challenges ahead, especially relating to the digital and economic divides between the two continents that characterise the cooperation landscape. Moreover, closing the existing digital and economic divides between developed and developing economies lends itself to joint bi-regional efforts. Given the potential dividends of STI, dialogue and negotiation across many themes and sectors between the two continents have given rise to a number of policy initiatives and funding schemes to facilitate bi-regional collaboration in joint endeavours. Apart from basic research, emphasis is increasingly put on the application of findings towards meeting major global challenges such as food and nutrition security, health and climate change.

References

African Union & European Union. (2007a). *The Africa-EU strategic partnership: A joint Africa-EU strategy.* Available from: http://www.africa-eu-partnership.org/sites/default/files/documents/eas2007_joint_strategy_en.pdf. Accessed 8 May 2017.

African Union & European Union. (2007b). *First action plan (2008–2010) for the implementation of the Africa-EU strategic partnership.* Available from: http://www.africa-eu-partnership.org/sites/default/files/documents/jaes_action_plan_2008-2010.pdf. Accessed 8 May 2017.

African Union & European Union. (2010). *Joint Africa-EU strategy: Action plan 2011–2013.* Available from: http://www.africa-eu-partnership.org/sites/default/files/documents/03-JAES_action_plan_en.pdf. Accessed 8 May 2017.

Barro, R. J., & Sala-i-Martin, X. (2004). *Economic growth* (2nd ed.). Cambridge: MIT Press.

Barugahara, I. N., & Tostensen, A. (2009a). *Science and technology for development: The institutional landscape in Africa and Europe.* Available from: https://caast-net-plus.org/object/document/83/attach/ScienceandTechnologyforDevelopment-1.pdf. Accessed 9 May 2017.

Barugahara, I. N., & Tostensen, A. (2009b). *Towards better synergy between S&T and development: Proposals and recommendations.* Available from: https://caast-net-plus.org/object/document/73/attach/2_2_1_Towards_Better_Synergy-2.pdf. Accessed 9 May 2017.

Booth, D., & Therkildsen, O. (2012). *The political economy of development in Africa: A joint statement from five research programmes.* Available from: https://differenttakeonafrica.files.wordpress.com/2012/04/joint-statement.pdf. Accessed 9 May 2017.

Court, J., Hovland, I., & Young, J. (Eds.). (2005). *Bridging research and policy in development: Evidence and the change process.* London: Overseas Development Institute.

Diyamett, B. D. (2008). *Scientific community, relationship between science and technology and the African predicament: Who is to blame and what can be done?* Available from: https://smartech.gatech.edu/bitstream/handle/1853/35621/Bitrina_Diyamett_Scientific_Community.pdf?sequence=1&isAllowed=y. Accessed 9 May 2017.

EU-Africa Summit. (2014). *Fourth EU-Africa summit 2–3 April 2014, Brussels, roadmap 2014–2017.* Available from: http://www.africa-eu-partnership.org/sites/default/files/documents/2014_04_01_4th_eu-africa_summit_roadmap_en.pdf. Accessed 9 May 2017.

European Commission. (2006). *The European consensus on development.* Available from: https://ec.europa.eu/europeaid/sites/devco/files/publication-the-european-consensus-on-development-200606_en.pdf. Accessed 9 May 2017.

European Commission. (2016a). *Press release: Joint communique between the African Union Commission and the European Commission at their 8th College to College Meeting.* Available from: http://europa.eu/rapid/press-release_STATEMENT-16-1301_de.htm. Accessed 9 May 2017.

European Commission. (2016b). *Press release: African Union Commission and European Commission meet to address shared EU-Africa challenges.* Available from: http://europa.eu/rapid/press-release_IP-16-1226_en.htm. Accessed 9 May 2017.

European Commission. (2016c). *Implementation of the strategy for international cooperation in research and innovation.* Available from: https://www.ffg.at/sites/default/files/downloads/progress_report_oct-2016.pdf. Accessed 8 June 2017.

European Commission. (2017). *Joint communication to the European Parliament and the Council for a renewed impetus of the Africa-EU partnership.* Available from: http://www.africa-eu-partnership.org/sites/default/files/documents/communication_for_a_renewed_impetus_of_the_africa-eu_partnership.pdf. Accessed 8 June 2017.

European Union. (2015). *Briefing. EU scientific cooperation with third countries.* European Parliamentary Research Service. Available from: http://www.europarl.europa.eu/RegData/etudes/BRIE/2015/564393/EPRS_BRI(2015)564393_EN.pdf. Accessed 26 June 2017.

Leach, M., Sumner, A., & Waldman, L. (2008). Discourse, dynamics and disquiet: Multiple knowledge in science, society and development. *Journal of International Development., 20*(6), 727–738.

NEPAD. (2014). *On the wings of Innovation: Science, technology and innovation for Africa 2024 Strategy (STISA-2024)*. Pretoria: NEPAD. Available from: http://www.hsrc.ac.za/en/events/seminars/science-tech-and-innovation-strategy. Accessed 16 May 2017.

NEPAD Office of Science and Technology. (2006). *Africa's science and technology: Consolidated plan of action (CPA)*. Available from: http://nepadwatercoe.org/wp-content/uploads/report_activities_cpa.pdf. Accessed 9 May 2017.

Romer, P. M. (2007). Economic growth. In *The concise encyclopaedia of economics*. Indianapolis: Liberty Fund.

Solow, R. M. (1956). A contribution to the theory of economic growth. *Quarterly Journal of Economics., 70*(1), 65–94.

Temple, J. (1999). The new growth evidence. *Journal of Economic Literature., 37*(1), 112–156.

United Nations. (2015). *Transforming our world: The 2030 Agenda for Sustainable Development*. Available from: https://sustainabledevelopment.un.org/content/documents/21252030%20Agenda%20for%20Sustainable%20Development%20web.pdf. Accessed 9 May 2017.

PART II

Cooperation in Food Security, Climate Change and Health

CHAPTER 3

The Dynamics of EU–Africa Research and Innovation Cooperation Programmes

Erika Kraemer-Mbula, Constantine Vaitsas, and George Owusu Essegbey

Abstract This chapter focuses on the practical achievements of existing Africa–Europe science, technology and innovation (STI) projects. It reviews six programmes that fund Africa–Europe STI cooperation, highlighting some of their successful cooperative projects, particularly in the fields of new water and sanitation technologies and green chemicals. This practical focus sheds light on the intrinsically unequal cooperation patterns among African countries. Participation of a diverse range of African partners, and private sector participation in Africa–EU STI cooperation, remain limited. The authors thus point out that future cooperation should focus on how to market and disseminate STI products and services.

E. Kraemer-Mbula (✉)
University of Johannesburg, Johannesburg, South Africa

C. Vaitsas
Forth/Praxi Network, Athens, Greece

G.O. Essegbey
Science and Technology Policy Research Institute of the Council for Scientific and Industrial Research, Accra, Ghana

A. Cherry et al. (eds.), *Africa-Europe Research and Innovation Cooperation*, https://doi.org/10.1007/978-3-319-69929-5_3

Keywords Framework programme • Horizon 2020 • European Union • African Union • Balanced cooperation • Co-financing • Private sector • Business networks

Introduction

The landscape of Africa–Europe collaboration in science, technology and innovation (STI) is becoming increasingly complex. Thematic areas are now addressing global, multi-sectoral concerns such as climate change; the conventional principles of "donorship" are being replaced by a growing search for equal partners and co-funding; and the need to address the global scientific divide and strengthen the STI capacities of low- and middle-income economies is more and more widely acknowledged. How have African countries and organisations, in particular, been affected by or contributed to these changes?

To answer this question, this chapter reviews past, present and future collaboration patterns of six funding programmes in which African and European countries and organisations have participated. These programmes are the European Commission's (EC) Framework Programmes (FP), with particular emphasis on the last two iterations, FP7 and Horizon 2020; the Cooperation in Science and Technology (COST) programme; Eurostars; the African, Caribbean and Pacific (ACP) countries' Science and Technology (S&T) programme; the European & Developing Countries Clinical Trials Partnership (EDCTP); and the African Union Research Grants (AURG) programme. Most of these programmes, with the exception of the AURG, are set in the context of the European Union (EU). As such, they should enable African partners to become important players in a bi-regional cooperation context. The chapter first highlights outstanding examples of successful and innovative projects funded under these programmes and then reflects on the way future collaborative relations can be strengthened.

The Seventh Framework Programme

The EU's first FP for research was introduced in 1984 to boost the scientific and economic development of the European Community, while promoting international cooperation. Its seventh iteration (FP7) was

Start date: 2007 - End date: 2013
Total budget: 50 billion euros
Funding body: European Union
In brief: FP7 was the key tool for the EU to respond to its needs in terms of employment and competitiveness. Its main objectives were to strengthen the scientific and technologic base of the European industry and to encourage international competitiveness.

Fig. 3.1 Overview of the Seventh Framework Programme (FP7)

launched in 2007. One of FP7's guiding ideas was that the promotion of the EU's strategic goals on research and development (R&D) was to be achieved through facilitated partnerships with third countries (i.e. countries outside of the EU, thus including African states) while addressing specific challenges that third countries face or which have a global impact (e.g. climate change). The programme had five major building blocks (see also Fig. 3.1):

- **Cooperation:** fostering collaborative research across Europe and other partner countries in different thematic areas such as health, food, agriculture and fisheries, nanoscience, environment and transport
- **Ideas**: supporting research on the basis of scientific excellence in areas including engineering, socio-economic sciences and the humanities
- **People**: supporting research mobility and career development both for the EU and internationally
- **Capacities**: strengthening the research capacities for the EU regarding research infrastructure, research potential, science in society and specific activities of international cooperation
- **Nuclear Research**: including research, technological development, international cooperation, dissemination of technical information and exploitation activities

For the first time, FP7 provided a broad opening for international cooperation in programmes and research schemes across the entire FP, while setting collaboration priorities with third countries and regions across the thematic work programmes. By defining specific actions for collaboration with third countries and regions in each of the thematic programmes, FP7 ensured that budgets for international cooperation were included at the level of each relevant call for proposals. Finally, the principle of partnership and dialogue was intensively applied in the specific actions for international cooperation with third countries and regions, in particular through the so-called INCO-NET instrument, of which CAAST-Net and CAAST-Net Plus are examples.

African participation in the FP7 rose dramatically. As is also reported in Chap. 2, 1315 participants from organisations in 45 African countries took part in 565 EU-funded projects, with a total budget of 178 million euros. In comparison, FP6 counted, in 2006, 882 African participants for 322 research projects, for an allocated budget of 95 million euros from the EU.[1] Under FP7, and as Fig. 3.3 also shows, South Africa, followed by Ghana, Uganda and Kenya, was the leading partner in terms of project participation.

Universities and governmental research institutions have been, so far, the leading project participants. In South Africa, the universities of Cape Town, KwaZulu-Natal and Pretoria, along with the Council for Scientific and Industrial Research, the Agricultural Research Council, the Institute of Natural Research Association and the National Research Foundation, were frequent FP7 participants. In Kenya, the University of Nairobi, the Ministry of Education, Science and Technology and the International Centre for Research in Agroforestry have been particularly active. Although fewer Ghanaian research organisations and universities participated in the programmes, the University of Ghana, the Kwame Nkrumah University of Science and Technology Kumasi and the Council for Scientific and Industrial Research were among the most active members. While some South African small- and medium-sized enterprises (SMEs) were involved, such as Research Africa, SME participation has remained limited to date, a trend that applies to most African countries (Fig. 3.2).

African FP7 participation received a new impetus with the 2010 Coordinated Call for Africa, also known as the "Africa Call", which placed African needs and priority research areas at the centre of the funding and programme design. It launched with the aim of addressing some of the S&T objectives of the Joint Africa–EU Strategy (JAES), in particular those of the Eighth Partnership on Science, Information Society and Space,

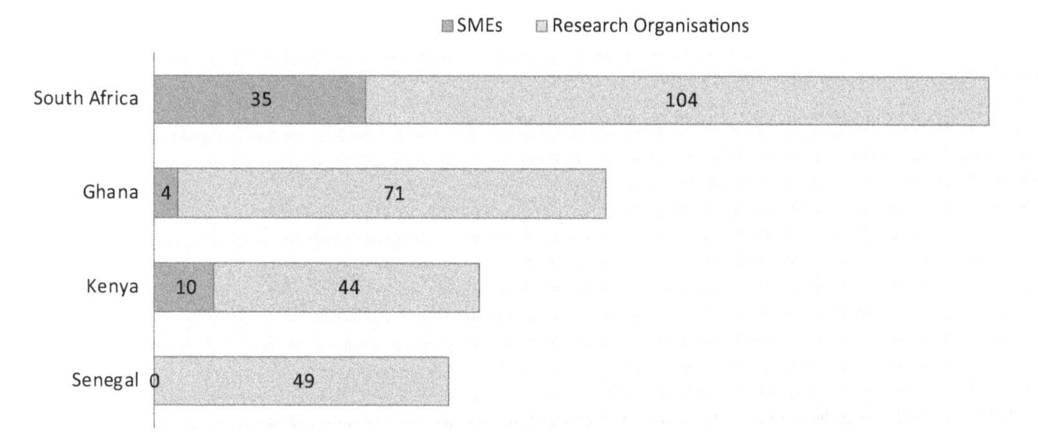

Fig. 3.2 SMEs and research organisations' participation in FP7 (Source: eCorda 2015)

seeking more particularly the joint elaboration of the Partnership's lighthouse projects by the African Union Commission (AUC) and the EC. The call funded 26 projects in three thematic areas: 15 on health, 7 on environment (including climate change), and 4 on food, agriculture fisheries and biotechnology. The funded projects intended to move away from a "donorship" approach to more equal partnership terms, combining the latest scientific discoveries with local knowledge to maximise research impact. Projects also aimed to strengthen local capacities in the relevant S&T fields and their applications, through training activities and the exchange of staff.

Again, South Africa was the most active country in the Africa Call, participating in 12 projects. In fact, South Africa's success under the FP7 was largely a result of a concerted effort to promote cooperation, undertaken by the European South African Science and Technology Advancement programme, an advisory, information and support platform for researchers, funded under FP7 and implemented by the South African Department of Science and Technology, as well as South Africa's FP7 Network of National Contact Points (NCPs) (EC 2009). As Fig. 3.3 shows, South Africa was followed by Tanzania (with 11 projects), Uganda (10) and Burkina Faso (10).

Within this context, the Water, Sanitation and Hygiene Technologies (WASHTech) project stands out for its contribution to the transfer of scientific knowledge through open access. It was a three-year action

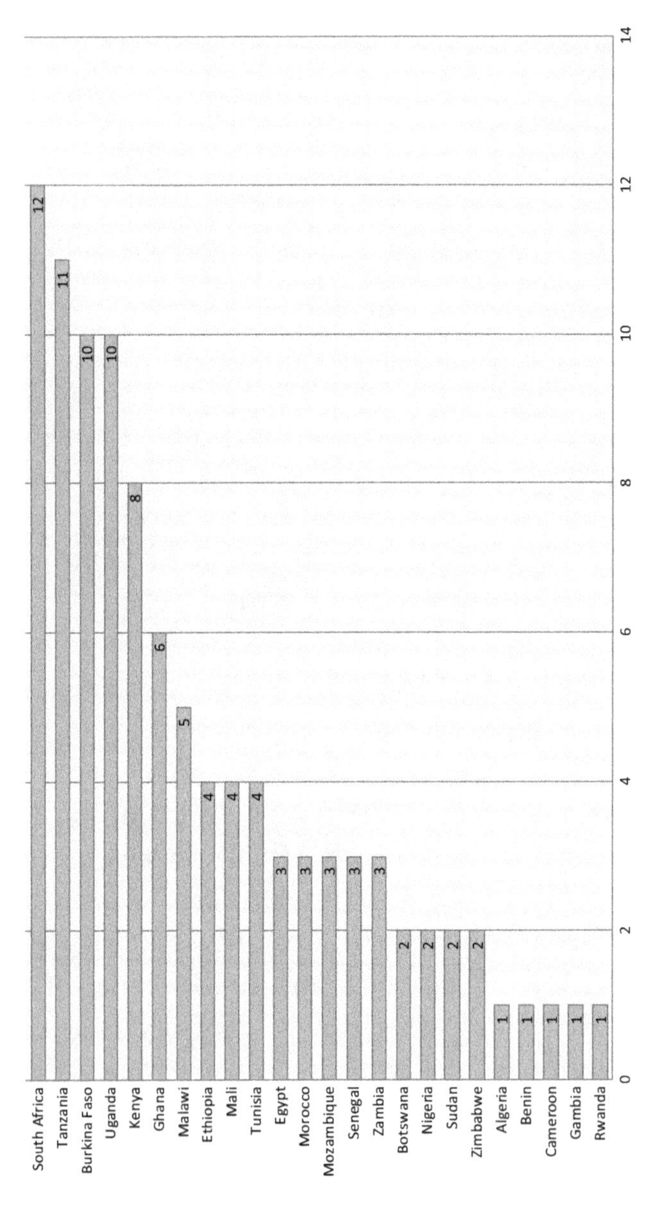

Fig. 3.3 Country participation in Africa Call projects (number of projects)

research initiative, spanning 2011–2013, backed by more than 2 million euros in funding, and involving African partners from Burkina Faso, Ghana and Uganda. The project was based on the premise that although the water and sanitation sectors are undergoing many changes, new technologies developed in these fields are simply not included in national strategies. The project resulted in two specific outputs: first, the development and introduction of an assessment tool, the Technology Applicability Framework, which provides a neutral approach for investigation of water, sanitation and hygiene (WASH) technological innovation; and second, the establishment of the Technology Introduction Process, which are multi-stakeholder and country-specific guidelines with agreed responsibilities to the successful introduction and uptake of WASH technologies in areas promising sustainable WASH service delivery. The project was set up to establish the required capacities, in order for these technologies to be incorporated in national policies. Moreover, WASHTech differentiated itself from other FP7 projects under the Africa Call by opening its outputs to validate water and sanitation technologies to the public domain. As such, it is an interesting case of non-commercial innovation, currently used in Europe and Africa and with potentially far-reaching impacts in both Europe and Africa.

Although FP7 offered an opportunity for African countries to collaborate with their European counterparts, the private sector showed limited participation. Considered to be the motor of innovation and technology diffusion in Africa's market-based economies, the private sector has the potential to bring co-funding opportunities in research cooperation and strengthen the sustainability of research networks. Future developments within bi-regional scientific collaborations have specifically sought to take this issue into consideration.

Horizon 2020

Succeeding FP7 in 2014, Horizon 2020 is the largest EU research and innovation (R&I) programme to date and is also expected to attract additional private investment funds based on the estimated results generated when entering the market. It has three pillars (Fig. 3.4):

Start date: 2014 - End date: 2020
Total budget: 80 billion euros
Funding body: European Commission
In brief: Horizon 2020 is the biggest EU Research and Innovation programme ever. Its main objective is to ensure Europe produces world-class science focusing on excellent science, industrial leadership and tackling societal challenges.

Fig. 3.4 Overview of Horizon 2020

- **Excellence of science** to reinforce and extend the EU science base
- **Industrial leadership** focusing on speeding up the R&D process behind new technologies and innovations that enable SMEs to grow
- **Societal challenges** reflecting the policy priorities of the EU strategy for 2020 and addressing major concerns by citizens in the EU and elsewhere (such as health and demographic change, food security, clean energy, transport, climate change and security)

Horizon 2020 is particularly geared towards acquiring additional financing through increasing the number of topics explicitly flagging "international collaboration", which increased from 12% of topics in FP7 to over 27% in the 2014–2017 round of funding calls. It also facilitates worldwide participation by reducing bureaucracy so that participants can focus on the substance of their R&I endeavours. Despite these efforts, the share of participation of partners from third countries in grant agreements for collaborative actions has fallen from 4.9% in the FP7 to 2.4% under Horizon 2020. As of October 2016, AU entities had 191 participations in 79 signed grants, receiving 31.2 million euros from the EU, while 2.9 million euros is from the non-EU budget. The equivalent first two years of FP7, however, counted with a larger African participation with 368 partners from 37 African countries involved.

As Figs. 3.5 and 3.6 show, South Africa still leads African participation in Horizon 2020 projects, as it did under FP7, followed by Ghana, Kenya and Uganda. The participation of the SMEs has remained, once again, limited.

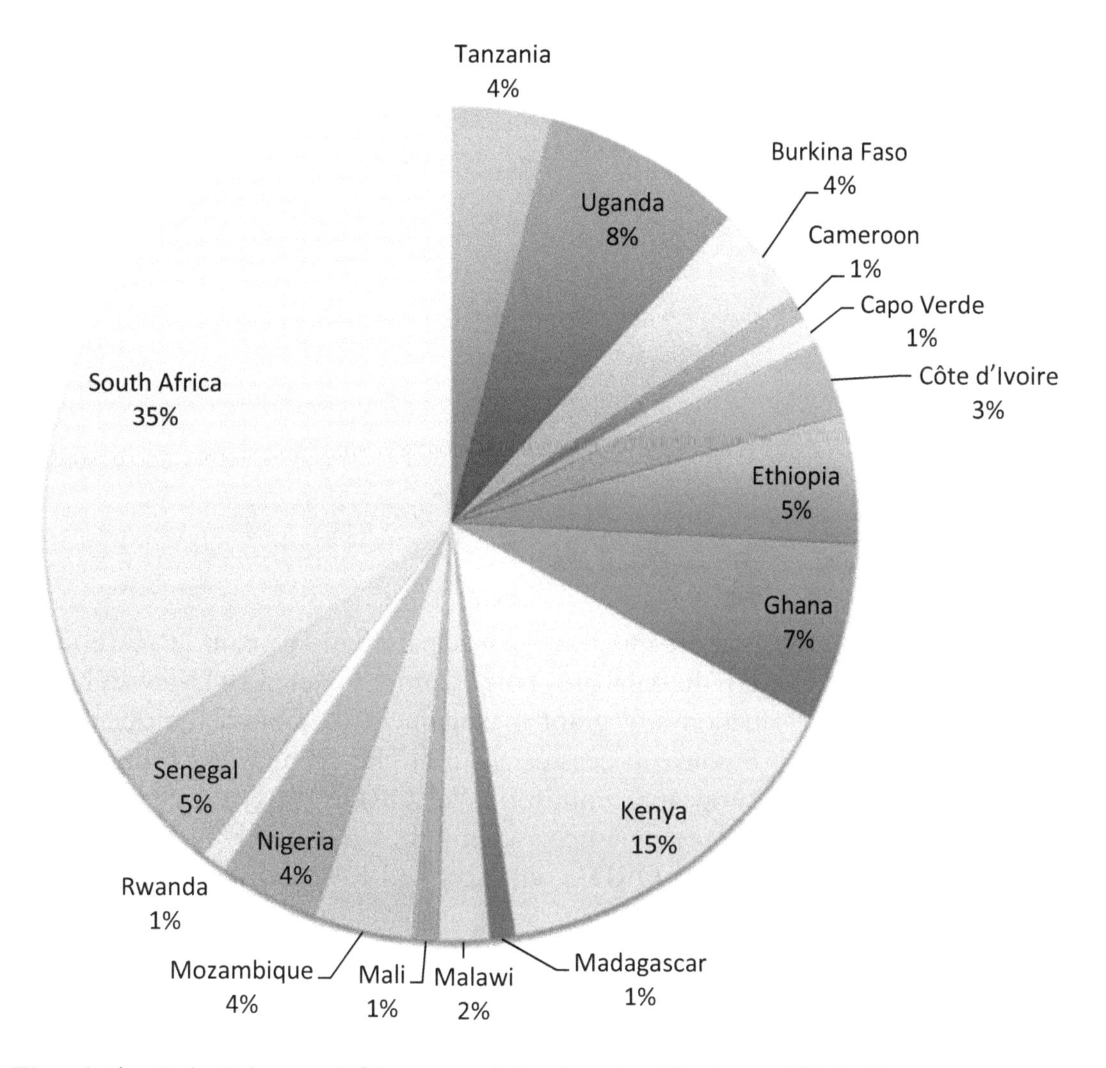

Fig. 3.5 Sub-Saharan African participation in Horizon 2020 (Source: eCorda, May 2016)

Most Sub-Saharan African organisations that answered the specific programme calls are research centres, institutions, universities and ministries. In the cases of South Africa, Ghana and Kenya, the most frequently participating organisations are the same noted in the case of FP7. In the case of Uganda, Makerere University is the leading participant, having joined already five different ongoing projects. Other frequent participants include the Uganda National Health Research Organisation and the National Agricultural Research Organisation.

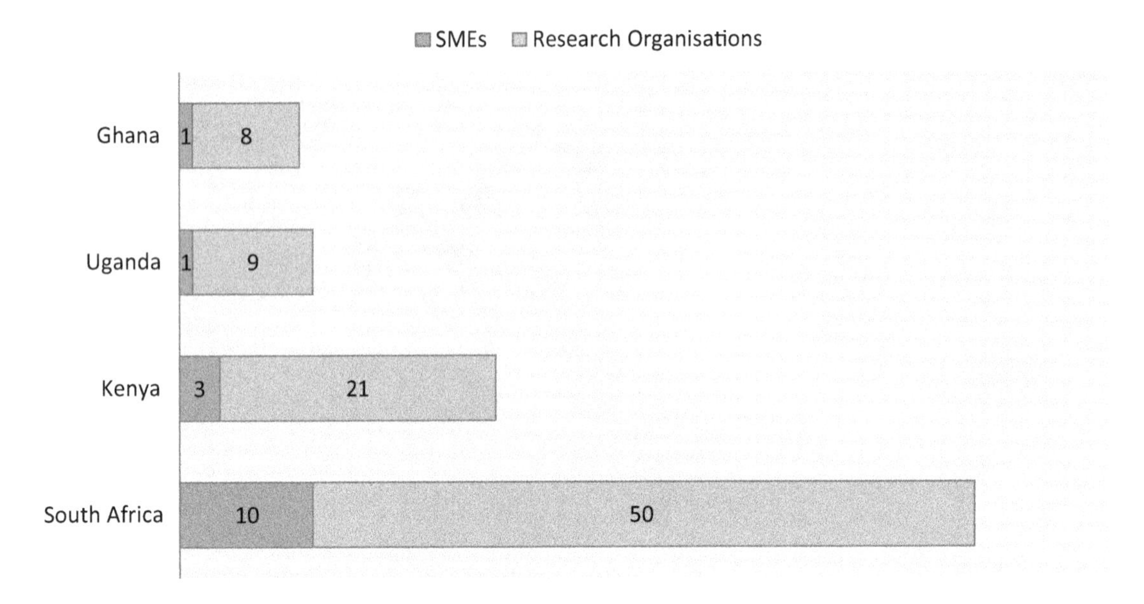

Fig. 3.6 SME and research organisations' participation in Horizon 2020 (Source: eCorda, May 2016; ECorda is the database of the "Common Research Datawarehouse, collecting proposals, evaluation and grant management data of all the operational systems automating key business processes around The Framework Programmes, H2020 – FP5" (Source: European Commission, "eCORDA and Cordis", November 2016, Available from http://www.ncpacademy.eu/wp-content/uploads/2016/11/20161103_eCORDA-and-CORDIS.pdf. Accessed 12 December 2017).)

Outcome Testimonial: Horizon 2020 information sessions for building bi-regional partnerships. *Compiled by Emeka Orji (National Office for Technology Acquisition and Promotion, Nigeria) and Melissa Plath (Finnish University Partnership for International Development, University of Jyväskylä).*

Access to information on research funding opportunities, particularly those from the EU, has often been difficult for researchers outside of Europe. EU funding can be a key instrument for supporting the formation and development of long-term, bi-regional research partnerships. To help address this asymmetry, CAAST-Net Plus organised a series of activities to disseminate information on opportunities for African participation in Horizon 2020, the EU's programme for funding research and innovation (2014–2020). For example, a Horizon 2020 workshop was held in Nigeria in 2013 to foster the participation of a wide spectrum of researchers.

Between 2013–2016, CAAST-Net Plus organised 15 workshops and information sessions on Horizon 2020, in 8 different African countries, with over 1100 participants from 28 different African countries. The events provided information on the relevant calls, rules, and strategies for successful proposals to Horizon 2020. Beyond providing information to the scientific community, CAAST-Net Plus also supported the nomination and training of an estimated 100 nominated or likely National Contact Points (NCPs) in Africa. NCPs are key disseminators of information on Horizon 2020: they can provide tailored advice and support to national researchers. CAAST-Net Plus organised five trainings for NCPs, inviting nominated or likely NCPs from across Africa.

As a result of these activities, and taking Nigeria as an example, researchers became more aware of the opportunities offered by Horizon 2020 and the potential of bi-regional partnerships in developing research proposals. Numerous Nigerian researchers requested to participate in various projects. While it is too early to observe specific or measurement outcomes *from this work, we do know that many of those who have attended the Horizon 2020 information sessions are actively engaged in proposals for bi-regional research grants, which is likely to result in a higher degree of African participation research and innovation projects, many under African leadership.*

In CAAST-Net Plus, we have experienced first-hand how more equal access to information and opportunities is important for building more—and more equal—bi-regional partnerships. Within our own partnership, we have noticed that when partners have equal access to information, they are more willing and able to contribute. We have also seen that supporting the nomination and training of NCPs contributes to knowledge sharing within a country. As disseminators of information, they play a key role in ensuring more equal access to information. CAAST-Net Plus has thus delivered a valuable contribution to the development of bi-regional research partnerships.

Since most projects funded through Horizon 2020 are still ongoing, it is premature to discuss the impact of Horizon 2020 on bi-regional cooperation. Instead, we introduce two current projects that reflect the type of programming work undertaken in Horizon 2020.

- *Africa–EU innovation alliance for water and climate (AfriAlliance).* The AfriAlliance project (2016–2021) is designed to foster collaboration between European and African actors in the fields of water innovation, research, policy and capacity development with a strategic objective to prepare African countries for future climate challenges. Although several initiatives and networks are active in this field, they remain fragmented. This project endeavours to connect and consolidate them and establish an overall coordination platform. The project has been allocated more than 3 million euros. Participation from Sub-Saharan Africa is significant, and includes partners from South Africa, Cote d'Ivoire, Ghana and Burkina Faso.

- *Integrated aquaculture based on sustainable water recirculating system for the Victoria Lake Basin (VicInAqua).* The VicInAqua project (2016–2019) aims at developing sustainable sanitation and recirculating aquaculture systems for the wastewater treatment and its reuse in agriculture in the Victoria Lake basin. The project tackles issues of food and health security while contributing to the protection of the ecosystems in Lake Victoria. VicInAqua is fully funded by the EC under Horizon 2020 with almost 3 million euros. Together with European participants, research associations and ministries from Uganda and Kenya will contribute to the development of a novel self-cleaning filter, which will be used in the aforementioned area.

Cooperation in Science and Technology (COST)

The European Economic Community and 19 European countries established the COST programme in 1971 to promote networks of researchers throughout Europe and beyond. Today, COST consists of 36 member countries, one cooperating state (Israel) and several international partner countries. Its objectives include capacity building through connecting high-quality scientific groups, offering networking opportunities for early stage researchers and increasing the impact of the current research among policymakers, regulatory bodies and the private sector (Fig. 3.7).

Although the programme does not fund research itself, it supports the creation of bottom-up networks of scientists and researchers, through the

Start date: 1971 – present
Annual budget per COST Action: average of 130.000 euros per Action
Funding body: European Commission and national budgets
In brief: COST is an intergovernmental organisation promoting the cooperation on science and technology research through creating networks, also known as "COST Actions". Those networks allow scientists to share their ideas with their peers, leading to proper dissemination of research and innovation inside and outside EU borders.

Fig. 3.7 Overview of the Cooperation in Science and Technology programme

so-called COST Actions. These are essentially networking tools to promote international coordination of nationally funded research and global cooperation. Typically, research must be relevant for at least five COST countries, while the financial support totals 130,000 euros per year for a four-year period—and must encourage global cooperation. The African countries already participating in COST Actions include Ethiopia, Mauritius, Namibia, South Africa and Sudan. Green chemistry, presented below, is a meaningful example of successful COST collaboration outside of the EU.

- *Cooperation between scientists from the United Kingdom and Ethiopia in the field of green chemistry* (COST 2017a). Green chemistry focuses on the design and use of less hazardous chemicals and processes. It has become particularly relevant in African countries, such as Ethiopia, as it offers unique opportunities to discover new biologically active molecules (for use in pharmaceuticals or agrochemicals) from the wide variety of flora endemic to the region. The collaboration of British and Ethiopian scientists produced a report comparing different methods for oil extraction in Ethiopia, researching milder chemicals for the extraction process. The findings were presented in major "green chemistry" conference organised in Germany in October 2004, and where the researchers from Ethiopia joined the COST Action D29 (COST 2017b). Through the COST Action, scientists were able to raise awareness about the field of green chemistry and to achieve a meaningful and efficient collaboration between the United Kingdom and Ethiopia. The project was considered a success: the report was widely acknowledged and adopted as teaching material from the New University of Lisbon (Engida et al. 2007). Most importantly, it gave visibility to indigenous chemical processes in Ethiopia that match the majority of the principles of green chemistry.

Eurostars

Together with EUREKA, an intergovernmental organisation for pan-European R&D funding and coordination, the EC established the Eurostars joint programme in 2008 (Fig. 3.8) (EUREKA 2017) to

Start date: 2008 – present Total budget: 1.4 billion euros
Funding body: EU (861 million euros) and national funding (287 million euros)
In brief: Eurostars is a joint programme between EUREKA and the European Commission, which supports international innovative projects in order to promote the development of innovative products, processes and services.

Fig. 3.8 Overview of the EUREKA programme

support international innovative projects led by R&D-driven SMEs, with the objective of bringing innovative products, processes and services to European and international markets. As it focuses on innovative, rapidly marketable products, processes and services, the selection process is highly competitive—there is no restriction on the technological area targeted; the only requirement is the clear aim to develop a new product, process or service. Any type of organisation can join a project consortium provided that the main partner is an R&D-driven SME.

While there was no African participation in the first Eurostars (2008–2014), its second iteration (2014–2020) saw South Africa become the only African country participating in EUREKA, signing in March 2016. Given its recent participation, the country is not yet active in any project under Eurostars calls. South Africa is nonetheless involved in two projects under ACQUEAU (EUREKA's cluster for water), described below.

- *The mine water as a resource (MINWARE)* project (2014–2017) aims to develop and demonstrate environmentally friendly solutions for mine wastewater from the mining and metal industries. The focus is on developing solutions for acidic metal-laden wastewaters in order to recover the valuable metals with new and less invasive methods and processes, together with the creation of viable economic solutions that are applicable worldwide. The project runs on a budget of 3 million euros, distributed within a consortium including research organisations, large technology providers, SMEs and their end users from Finland, South Africa and Sweden. South Africa's role is to foster the involvement of South African SMEs as end users, specialised in designing and developing biological methods for wastewater treatment.

- *A long-term, sustainable treatment option for acid mine drainage (AMD)*, the *VitaSOFT* project, implemented over a 30-month period, focused on the demonstration of a treatment process for AMD. Under the coordination of Vitaone8, a South African SME specialised in developing innovative water treatment technologies, and participants from the United Kingdom and South Africa focused on the successful demonstration of the VitaSOFT process. VitaSOFT is an active biological sulphate-reducing process, designed to reduce the volume of solid waste and provides the potential to recover valuable by-products. The process can also be used for the disposal of other waste such as industrial biodegradable organic waste, which would reduce the cost and the risk that companies take to dispose of waste themselves. The project was financed under the ACQUEAU RTD project call, with a 1 million euros grant co-funded by the EU, South Africa and the United Kingdom.

African, Caribbean and Pacific Countries' Science and Technology (S&T) Programme

Funded by the EU and implemented by the ACP secretariat, the ACP S&T programme responds to the need for joint and systemic approaches in support of STI. The programme acknowledges a direct link between building and enhancing strong S&T capacities to support research, development and innovation in the ACP region, and the identification and formulation of activities, processes and policies critical to sustainable development. Its first iteration, launched in 2008, funded 36 projects (out of 200 proposals), ranging from quality health care, environmental research activities, energy, transport, agriculture and agro-industry as well as sustainable trade, which received a total of 58 million euros. Its second iteration, launched in 2013, funded 21 projects focusing more specifically on energy and agriculture with a total budget of 20.8 million euros (Fig. 3.9).

These successive ACP S&T programmes aim at reducing the scientific and technological divide between ACP countries and the most industrialised countries while increasing partners' ability to better assess their research needs, build stronger networks and implement research politics; strengthening the STI capacities for ACP countries to create, update and

Starting date: 2008 – present
Total budget: 78 million euros
Funding body: European Union
In brief: ACP S&T Programme is a cooperation programme between the EU and the ACP countries, focusing on the enhancement of the internal capacity in S&T of ACP countries to support research, development and innovation in their regions.

Fig. 3.9 Overview of the African, Caribbean and Pacific countries' science and technology programme

use scientific knowledge; and enhancing the use of STI as a key enabler for poverty reduction, growth and socio-economic development. The funded projects are expected to establish or strengthen networks between ACP countries, but also to create global links. To successfully address its goals, the programme is designed to support the efforts of ACP countries on three levels (1) institutional, administrative and policy-making; (2) academic, research and technology; and (3) business and civil society. Two successful examples of such projects supported under the ACP S&T programmes are described below.

- *Western Africa biowastes for energy and fertiliser* (WABEF) is a research-development and capacity building project focused on the most effective ways to recycle organic residues and produce energy and fertilisers in West Africa. Waste management is a priority in the region, mainly due to the rapid population growth and high urbanisation rates. Therefore, this initiative supported the development of practices and technologies that prioritise both the recycling but also the repurposing of organic waste in the region. Using existing knowledge, the project aimed to devise tools in a participatory way, involving executives and public decision makers, NGOs and teachers, municipalities and agribusinesses, in order to assess the process of anaerobic digestion as the most applicable process for West Africa's waste treatment policy. The project was launched in 2014 for three years, receiving more than 700,000 euros from the EU. Participation from African countries included organisations from Senegal, Benin and Mali.

- Strengthening capacities and informing policies for developing value chains of neglected and underutilised crops in Africa. This project focused on the research into neglected or underused species and the development of national action plans in order to incorporate them in agricultural systems in Eastern, Western and Southern Africa. Neglected or underused plant species offer niche markets and incomes for poor farmers, as well as options for climate change adaptation. However, their potential is often neglected, mainly due to globalisation, population growth and urbanisation. The project aimed at changing this situation, by involving local stakeholders such as farmers, processors, researchers and the private sector in innovation platforms to upgrade the value chains of such crops. The project ran from 2014 to 2017 and received almost 1 million euros from the EU. Participation from African countries included research organisations and universities in Kenya, Benin and Zimbabwe.

European & Developing Countries Clinical Trials Partnership (EDCTP)

The EDCTP programme was launched in 2003 as a European response to the need to enhance research collaborations between scientists and accelerate the clinical development for new or improved solutions to tackle poverty-related or neglected diseases such as HIV/AIDS, tuberculosis and malaria (see also Chap. 6). All projects funded under the EDCTP are implemented by partnerships between African and European research institutions in collaboration with the pharmaceutical industry. The EDCTP Association has members from 14 African countries (namely Burkina Faso, Cameroon, Congo, Gabon, The Gambia, Ghana, Mali, Mozambique, Niger, Senegal, South Africa, Tanzania, Uganda and Zambia) and 14 European countries (Austria, Denmark, Finland, France, Germany, Ireland, Italy, Luxembourg, Netherlands, Norway, Portugal, Spain, Sweden and the United Kingdom); it also remains opened to new memberships (Fig. 3.10).

The first EDCTP (2003–2015) received 378 million euros from the EU and European countries that are members of the EDCTP Association. It funded 246 projects for a total cost of 212 million euros (see EDCTP 2014). The second programme (2014–2024) received 700 million

Start date: 2003 – End date: 2024
Total budget: 378 million euros (for the 1st programme) 683 million euros (for the 2nd programme)
Funding body: European Union and European member states
In brief: EDCTP is a public-private partnership between the EU, Sub-Saharan Africa countries and European countries, which focuses on the development of new or improved. interventions for the prevention or treatment of several diseases in Sub-Saharan Africa.

Fig. 3.10 Overview of the European & Developing Countries Clinical Trials Partnership

euros from European countries and Horizon 2020. The ENDORSE project, described below, is one of the project in the second EDCTP programme.

- *Enhancing individual and institutional infectious disease outbreaks response capacities of healthcare professionals to mitigate infectious emergencies in the Northern Uganda region (ENDORSE).* The project (2016–2017) provides training to healthcare workers in biosafety and protection, against infectious diseases, and focuses on the region of Northern Uganda, where capacity building will benefit healthcare workers in both laboratories and patient-care settings. The project addresses the existing gaps in human resources for health, the wide disparities in health status across the country and the weakness of capacity in planning, management and human resource development. To achieve these objectives, a sustainable Train-the-Trainer model will be implemented and tested through training phases. Uganda is the only participant from Sub-Saharan Africa, joining European participants from Italy and Ireland. ENDORSE is funded with almost 200,000 euros by EDCTP.

The African Union Research Grants Programme

The AURG was initiated by the Department of Human Resources, Science and Technology (DHRST) of the AUC to support a pan-African R&D programme through grants and direct funding. In line with the African

Start date: 2014 – End date: 2020
Total budget: 17.5 million euros allocated in the second phase
Funding body: European Union
In brief: The AURG programme supports Pan African research and development through grants and direct funding, as tools for sustainable development, as well as building and strengthening Africa's S&T capacities.

Fig. 3.11 Overview of the African Union Research Grants programme

Union's Science Technology and Innovation Strategy (STISA-2024) and with the JAES and the Priority 3 on human development of the EU–Africa partnership, the AURG programme supports collaborative research and R&I activities contributing to the sustainable development of African countries. A key priority is to develop the capacity of the AUC to design, implement and monitor R&I funding programmes, to establish the basis for a credible and reputable African framework programme for R&I (to attract additional funds from other sources such as AU member states and other partners and donors), and to enhance intra-regional and North-South scientific research consortia (Fig. 3.11).

Already in its second phase, the AURG programme is financed through a financial agreement between the EC and the AUC under the pan-African programme (2014–2020), whereby the EC has allocated a budget of 17.5 million euros for two calls in 2016 and 2017. The call launched in 2016 awarded grants to research projects addressing the priorities set out in the R&I Roadmap on Food & Nutrition Security and Sustainable Agriculture, namely eradicating hunger and ensuring food and nutrition security, which were determined through the EU–Africa High Level Policy Dialogue on STI.

Africa–Europe STI Cooperation Programming: A Look Forward

The thematic areas in which Africa–Europe cooperation on STI takes place have expanded both qualitatively and quantitatively over the years. Furthermore, the link between research areas and priorities, funding instruments, and joint strategies increasingly reflect common values and objectives as opposed to the EU's one-sided agenda setting of the past. Nevertheless, this dynamic and rich landscape is not without challenges: the private sector is still under-represented and Africa-wide participation in bi-regional programming is still uneven.

Although most of the research funding and programming described above are open for private sector participation, its involvement remains minimal, and African participation in cooperation projects remains limited to a few public universities and research organisations. The long-term impact of applied research activities may be limited, if they are not coordinated with commercial actors who are interested to develop new processes, products and services. Low levels of private sector participation can be explained by a limited level of awareness among private sector actors regarding opportunities under FP7 and Horizon 2020; the subsequent lack of effective response to the calls proposals; and the lack of strategic alliances with European institutions and consortia.

Outcome Testimonial: Strengthening African capacities in collaborative relations with the EU within STI. *Compiled by Jean Albergel, Mamohloding Tlhagale, Johan Viljoen (Institut de Recherche pour le Développement, IRD) and Toto Matshediso (South African Department of Science and Technology).*

ERAfrica was launched in early 2011 as the result of a number of European and African countries being eager to better coordinate and strengthen their individual bilateral collaborative relations in STI. The project was aimed at helping to realise the first action plan of the JAES, seeking in particular to strengthen African capacities in STI. The concept of such a project was expressed during a CAAST-Net stakeholders' meeting in Mombasa, Kenya, in November 2009. The consortium in charge of the ambitious ERAfrica project was mainly composed of partners from CAAST-Net.

Through ERAfrica, funding parties from 15 African and European countries jointly created the necessary funding mechanisms and processes leading up to a first call for research proposals in which partners participated on an equal footing. The joint call for projects covered three thematic fields: renewable energies, interfacing challenges and new ideas. Project could involve three types of collaborative activities including collaborative research, collaborative innovation and capacity building. With a total of 10.7 million euros available for funding, the call generated 124 proposals from which ERAfrica selected 17 projects to be funded. The selected projects represent a total amount of 8.29 million euros and 65 institutions (31 in Africa) from 18 countries (8 African countries), working together in these projects.

As 9 African institutions and 8 European institutions have the important role of project coordination, the overall picture shows that ERAfrica indeed lives up to its aim of "true" partnerships and enhanced African capacities within research collaboration. These ERAfrica-funded projects are good examples of how research, development and innovation can be used to improve the lives of the African and European citizens and particularly in the field of health systems, food and nutritional security. At the end of the first phase of ERAfrica, there was a great desire among the funding parties to see it continue even if it would be without funding from the EC.

A number of instruments exist to enhance and facilitate international collaboration and promote the internationalisation of non-European SMEs. If properly and more widely implemented, they could bring significant improvements to international cooperation in R&I. The Business Cooperation Centres (BCCs), established through the initiative of the Enterprise Europe Network (EEN) in major international growth markets (such as Brazil, Russia, China and India) is one of them. The BCCs serve as a contact point for EU SMEs to enter international markets and to establish connections with local firms. In Sub-Saharan Africa, Cameroon and Nigeria have already established BCCs in order to facilitate business, technology as well as research partnerships between local SMEs and European SMEs based on common interests and a desire for mutual benefits (EEN 2016). More recently, the Enterprises of Cameroon were established in Douala. They are a professional association with expertise on how to facilitate cross-border commercial cooperation. Likewise, Nigeria established the Nigerian–Belgian Chamber of Commerce in Lagos, with the objective of promoting collaboration between Nigerian and Belgian enterprises by creating a platform and a friendly environment to support business development. BCCs can provide an effective platform for SMEs to internationalise their business and to explore R&I partnerships.

The NCPs network is another tool created to improve the implementation of the funding instruments of the EU. The EC considers NCPs to be vital partners for the implementation of funding programmes such as Horizon 2020. An NCP is a trained individual, officially appointed by its host organisation (upon EC endorsement) whose mission is to guide his/her country's participation in EU-funded programmes. NCPs serve as a source of strategic information. They advise SMEs and other organisations on how to access European consortia, through acquiring knowledge of the latest developments in R&I and gaining access to a pool of international partners for future collaboration. Furthermore, NCPs provide free of charge support in the applicant's own language.

Countries with well-established NCPs usually increase their participation in European programmes, through raised awareness of programmes, specific calls and technical requirements. Additionally, NCP networks help to identify pockets of excellence in the given country, ensuring their alignment with specific calls. In the case of Sub-Saharan Africa, there is evidence that the NCP network has been effective in enhancing participation in the calls for proposals in FP7. Indeed, the evolution of NCPs seems to be positively related to the country's increased participation in FP7 projects.[2] It is worth noting that other African countries have also increased their participation without an established NCP network, which can be explained by effective informal networks of researchers collectively applying to EU funding.

Conclusion

Although African countries have successfully participated in several EU funding programmes, African participation within Horizon 2020 appears to have decreased in comparison to FP7. At the same time, we see a wider range African countries applying to other funding schemes. It is evident that more work needs to be done to engage the private sector. This process must begin with unravelling the multiple limitations and barriers that private sector organisations experience in becoming aware of funding programmes, and how to access them. Experience has shown that there is limited understanding of the approaches and tools for enhancing innovation cooperation. Some existing tools have proven to be effective, and it would be useful to build on their experience. Networks such as the EEN can provide a great platform to strengthen the collaboration among companies and research organisations between Europe and Africa.

Bi-regional STI collaboration is increasingly becoming more reflective of the political aspiration towards co-ownership and equal partnership for mutual interest and mutual benefit. Focusing on areas of common interest and sharing common values are key ingredients of co-owning projects, and these are areas where significant improvements have been achieved. However, more remains to be done in terms of co-financing. The review of funding programmes in this chapter points out that bi-regional cooperation remains largely dependent on European funding. There has been some experience and experimentation with co-financing models showing excellent results, as exemplified by the ERAfrica consortium (under EU FP7), and the EDCTP. The recently established AURG programme, although still dependent on EU funding, is gradually building African institutional capacity to manage pan-African research programmes. Such programmes, eventually fully funded and owned by African countries, would expand the space for cooperation with international partners, including the EU, and would provide a richer and more diverse basis for bi-regional cooperation.

Notes

1. It must be noted that FP7, which spanned the period 2007–2013, had a longer duration than the FP6 (2002–2006). This may account for a proportion of the difference in total participation by and total financial allocation to African participants.
2. The efforts of CAAST-Net and CAAST-Net Plus have increased the NCP numbers in African countries. Through organising workshops and information days both on NCPs and Horizon 2020 the partners of CAAST-Net Plus have reached several countries in the Sub-Saharan region including Sudan, Cameroon, Angola, Mozambique, Mauritius, Uganda, Malawi, Tanzania, Kenya, Nigeria and Ghana (CAAST-Net 2013).

References

CAAST-Net. (2009). *Africa-Europe cooperation in science and technology: Status and way forward 10–11 November 2009 Mombasa, Kenya (summary report and recommendations)*. Available from: https://caast-net-plus.org/object/document/1285/attach/CAAST-Net_2009_Stakeholders__Conference_Conclusions.pdf. Accessed 18 May 2016.

CAAST-Net Plus. (2013). *Status of African national contact points.* Available from: https://caast-net-plus.org/object/document/642/attach/D5_2_4_1_African_NCP_status_report.pdf. Accessed 22 July 2016.

COST. (2017a). *A new COST success story: Green chemistry and cooperation with African countries. Available from:* http://www.cost.eu/media/newsroom/node_751. Accessed 27 June 2017.

COST. (2017b). *CMST COST action D29 sustainable/green chemistry and chemical technology*. Available from: http://www.cost.eu/COST_Actions/cmst/D29. Accessed 27 June 2017.

EDCTP. (2014). *Assessment of the performance and impact of the first programme of the European & developing countries clinical trials partnership (EDCTP)*. Available from: http://www.edctp.org/web/app/uploads/2015/03/Assessment-of-the-performance-and-impact-of-the-first-EDCTP-Programme_Technopolis-Group_18SEP2014.pdf. Accessed 27 June 2017.

Engida, T., Nigist, A., Licence, P., & Poliakoff, M. (2007). *Empowering green chemists in Ethiopia*. Available from http://science.sciencemag.org/content/316/5833/1849.full. Accessed 27 June 2017.

Enterprise Europe Network. (2016). *A network outpost in sub-Saharan Africa*. Available from: http://een.ec.europa.eu/news/news/network-outpost-sub-saharan-africa Accessed 18 May 2016.

EUREKA. (2017). *About Eureka*. Available from: http://www.eurekanetwork.org/about-eureka. Accessed 27 June 2017.

European Commission. (2009). *International cooperation with Africa in FP6*. Brussels: European Commission.

CHAPTER 4

Bi-regional Scientific Cooperation on Food and Nutrition Security and Sustainable Agriculture

Jean Albergel, Arlène Alpha, Nouhou Diaby, Judith-Ann Francis, Jacques Lançon, Jean-Michel Sers, and Johan Viljoen

Abstract This chapter argues that Africa–Europe cooperation still faces two significant challenges: first, the practical implementation of innovative solutions to the challenge of assuring food security is still lagging behind;

J. Albergel (✉) • J. Viljoen
IRD-CNRS-CIRAD Joint Office In Pretoria, Institut de Recherche pour le Développement (IRD), The Innovation Hub, Pretoria, South Africa

A. Alpha • J. Lançon • J.-M. Sers
Centre de Coopération Internationale en Recherche Agronomique pour le Développement (CIRAD), Paris, France

N. Diaby
Université Cheikh Anta Diop, Dakar, Senegal

J.-A. Francis
Technical Centre for Agricultural and Rural Cooperation ACP-EU, Wageningen, The Netherlands

A. Cherry et al. (eds.), *Africa-Europe Research and Innovation Cooperation*, https://doi.org/10.1007/978-3-319-69929-5_4

second, equal partnerships still need to be further institutionalised in order to become more enabling of positive change. As food and nutrition security issues touch on structural socio-economic, political and even environmental constraints, they require extensive networks of research, innovation and institutional collaboration. Despite several achievements, persisting asymmetries continue to burden the achievement of food and nutrition security goals in Sub-Saharan Africa. The authors point out the strategic role that African countries could play first in cooperating with less developed European countries and in promoting alternative concepts of nutrition and environmental development on the global stage.

Keywords Innovative solution • Equal benefits • Global development • Harmonised resources • Geographical representation • Agricultural platforms • Poverty • Health • Productivity

Introduction

Globalisation has changed the way knowledge is produced, shared and used. Major global challenges such as climate change, poverty, infectious disease, threats to energy, food and water supply, security and the digital divide highlight the need for effective global science, technology and innovation (STI) cooperation to promote sustainable development, notably in the developing world (European Commission 2009). In Africa, governments have recognised the importance of STI for this purpose, and as a result, the African Ministerial Conference on Science and Technology adopted the Consolidated Plan of Action (CPA) on STI in 2005 (see also Chap. 2). The CPA has since been revised as the Science, Technology and Innovation Strategy for Africa (STISA-2024), endorsed by African Heads of State and Government in July 2014 (NEPAD 2014).

Eradicating hunger and achieving food and nutrition security and sustainable agriculture (FNSSA) is one of the six priority areas of STISA-2024, while strengthening international cooperation is identified as one of the mechanisms for implementing actions in pursuit of using STI for socio-economic development and growth on the continent. In this way, the strategy is not only well aligned to its European counterpart (the Common Agricultural Policy) as well as placed squarely within the Joint Africa–EU Strategy (JAES) for region-to-region scientific cooperation and partnership. It is also fully a part of an increasing drive towards ownership of the agricultural science agenda by African countries themselves. This drive is championed by, among others, the Forum for Agricultural Research in Africa (FARA) and articulated in the document Science Agenda for Agriculture in Africa—also known as S3A (FARA 2013)—which was similarly endorsed by African leaders in 2014. Set against this policy backdrop, this chapter critically discusses the relationship between Africa and Europe in the domain of FNSSA. It highlights the extent to which FNSSA has featured within bi-regional STI cooperation more broadly, and it identifies critical success factors for FNSSA project partnerships.

Situational Analysis

The idea that agriculture in Africa is too important to be outsourced has led to the creation of several platforms operating at continental and sub-regional level and, similar to FARA, aiming at encouraging African countries to invest in sufficient scientific capacity to support agricultural transformation. Examples of platforms operating at continental and sub-regional level include the Association for Strengthening Agricultural Research in Eastern and Central Africa (ASARECA), the Conseil Ouest et Centre Africain pour la Recherche et le Développement Agricoles/West and Central African Council for Agricultural Research and Development (CORAF/WECARD) and the Centre for Coordination of Agricultural research and Development for Southern Africa (CCARDESA). These platforms coordinate the implementation of such programmes by facilitating collaboration among stakeholders and carrying out capacity building initiatives. Additional key functions of these platforms include knowledge management and dissemination, as well providing direct inputs into policymaking organs at national, regional, continental and international levels, including the African Union Commission (AUC), the New Partnership for Africa's Development Agency (NEPAD), the European Union (EU) and the World Bank.

As the preceding discussion points out, many well-documented and clearly articulated agricultural programmes and policies have been developed to address FNSSA in Sub-Saharan Africa. In addition, many African governments, regional bodies and organisations have been deeply involved in implementing strategic food policies and exploring research priorities. As previous chapters have noted, the need for cooperation at national, regional and international level is widely acknowledged, both politically and within the associated policy frameworks. Similarly, the need to increase food supply by raising production capacities, harnessing trade and improving natural resources management has repeatedly been emphasised. The pan-African policy framework established by the United

Nations, the Comprehensive Africa Agriculture Development Programme (CAADP) and NEPAD's Framework for African Food Security (FAFS) also recognised the need for a better application and optimisation of new technologies, and for improving the diversity and quality of diets.

Priorities for FNSSA in Sub-Saharan Africa are much broader than just increasing availability: poverty, food insecurity, poor health and malnutrition are interrelated issues also affected by the lack of political stability, environmental degradation and limited technical capacities. All these areas impact food productivity and are intended to be addressed by other cooperation programmes. Yet, despite the elaboration of most of these policies and programmes, poverty, hunger and malnutrition are still high in African countries: the UN Food and Agriculture Organization (FAO) estimates that, in 2014–2016, 233 million people in Sub-Saharan Africa were hungry/undernourished. The FNSSA goal remains to be achieved. As with other such intractable issues or wicked problems, such failure suggests a missing link between research outputs and FNSSA realities.

Outcome Testimonial: Increasing access by Beninese small- and medium-sized enterprises' (SMEs) to global markets by improving the quality of food products. *Compiled by Andrea Cefis (Belgian Development Agency, Benin).*

As a result of the "shrimp crisis", a food safety scare in 2002, Benin banned the export of the shellfish to Europe to avoid international sanctions as a response to inadequate food safety control systems operating in the country. This situation had a significant negative economic impact on Beninese industry and exports. In 2012, the European Commission (EC) and the Government of the Republic of Benin, in collaboration with the Belgian Development Agency, provided a 2.8 million euros grant to initiate the "Improving Food Safety" project. Running until May 2017, the objective of this Africa–Europe collaboration was to develop an adequate food safety control system, thereby helping SMEs to improve the quality of their food products and, at the same time, to strengthen the competitiveness of the country's products in international markets.

The project operated on three levels. First, the Beninese Food Safety Agency used it to refine a food security policy based on food safety risk analyses, and to design an improved food control system for assessing the adequacy of food safety. Second, the Laboratory for Control of Sanitary Food Safety (LCSSA) used the project to strengthen the analytical capacity of it research staff. As a result, in March 2016 the LCSSA was accredited according to the international ISO 17025 standard, allowing Benin to boast an approved international laboratory and enabling private industries to perform globally accepted product analyses locally before exportation. Third, on the private sector level, the project supported agri-food industries by training staff of SMEs to implement food hygiene initiatives, including Hazard Analysis Critical Control Point (HACCP).

Positive outcomes of the project are already visible: 18 Beninese SMEs have implemented HACCP, allowing them to export their products to Europe and the United States—for instance, cashew producers have obtained contracts with American enterprises and pineapple juice producers with French customers. Local producers of spirulina (blue-green algae used in numerous food products) now sell their goods to international institutions such as FAO and WHO (World Health Organization) to combat malnutrition. Furthermore, the Improving Food Safety project helped other agri-food industries, such as pineapple production, to develop food hygiene policies, while similar initiatives are now supporting agri-food industries to achieve conformity to international food hygiene standards such as ISO 22000, GLOBALGAP and ECOCERT.

The Realisation of FNSSA Priorities in Africa–Europe STI Cooperation

Europe–Africa STI collaboration has a long history that can in part be traced to 1983, when the EU's international cooperation on research programming commenced. Its benefits have been confirmed in more recent years by the increasing number of joint projects (including on FNSSA), their budgetary allocations, and the number of participating organisations involved (see also Chap. 3). More specifically, the EU has been instrumental in supporting continental and sub-regional research coordination platforms dealing with FNSSA, such as FARA and ASARECA, as well as Africa–EU bi-regional platforms such as PAEPARD (The Platform for African European Partnership on Agricultural Research for Development). In 2007, the JAES was adopted in response to new geopolitical changes, globalisation and the processes of integration on the two continents: it was the expression of an overtly political partnership that distinguished itself from the previous Africa–Europe policy initiatives by pointing out the need to address joint priorities for a more egalitarian and mutually beneficial cooperation.

In this context, the positive contributions of scientific and technological research, development and innovation together with the acknowledged role of research capacity for economic and social growth, as well as poverty alleviation have become explicit—in particular for building knowledge-based societies and addressing global societal challenges of mutual interest. The High Level Policy Dialogue (HLPD), which sets the agenda for the EU–Africa STI partnership and oversees progress, is also a forum for sharing and disseminating ideas to inform development policies at national and regional level.

Although FNSSA remains high on the list of priorities for Africa–Europe collaboration, the extent to which this collaboration has been successful in responding to the issue is in question. Partial progress is certainly undeniable: FNSSA is no longer limited to agricultural issues, as nutrition has become increasingly important. Indeed, several projects under Horizon 2020 such as LEAP-AGRI (see below) have been launched as part of the Africa–EU FNSSA partnership under the JAES. Whereas the "key" issues were previously restricted to producing bigger crops through more intensive growing practices, FNSSA is now considered much more broadly and includes, in relation to the sustainability of production and transformation systems, their impact on livelihood and ecosystem services. Other concerns, such as how to add value and create jobs, the efficiency of production models (large *versus* small-scale farming), access to market and entrepreneurship, and the food system as a whole have also become mainstreamed.

The JAES action plan insufficiently addressed the FNSSA priorities outlined in CAADP and in pillars III and IV specifically. However, the issue has more recently secured greater attention and support, and manifested in a bi-regional research and innovation (R&I) partnership on FNSSA established in 2016. Significant challenges remain though in ensuring that all the available knowledge is used to inform policy, improve food systems and processes, expand product range, markets and trade and support innovation for social and economic gain in both Europe and Africa. Indeed, only a few projects of the CAADP-FAFS actually address food stability as a priority issue; and private sector involvement in EU research framework programmes is poorly represented (about 15.5% of the participants) and so is civil society (only 1.5% of participating organisations). A major barrier to private sector engagement has been the differing motivations of business enterprises and research institutions, and the limited follow-through on research outputs after projects ended. The transfer and dissemination of knowledge also remains limited, as farmers often do not see the direct benefits of research cooperation projects. All of this suggests a dearth of mechanisms for making the knowledge available, understandable and convincing.

Future engagement in this area should focus on mechanisms to improve the accessibility of the outputs of joint Africa–Europe research cooperation as well as on making the knowledge accessible to a wider public. In addition, the enhanced capacity and knowledge created through cooperation should be used to improve STI, agriculture and FNSSA policy processes on both continents as well as to bring about greater synergy among the various policy instruments and implementing agencies. While this may be occurring to some extent already, ensuring greater continental, regional and national ownership of the FNSSA research and policy agendas, and developing a more robust research infrastructure (particularly in Africa) is sorely needed.

Critical Factors for Better and More Frequent Cooperation

A number of policy and practical programmes exist to influence bi-regional research cooperation in FNSSA. Yet, more can be done to link research outputs to implementing or "spending" organisations, including government agencies, in order to ensure that research outputs lead to clear *outcomes* (CAAST-Net Plus 2016). There is a case for arguing that openness to fair international cooperation should be standardised within research teams, while intellectual property agreements (where relevant) and the funding of research exchanges should become the responsibility of research organisations. Standardising, and/or clarifying, visa regulations is one example of how international research exchanges can be facilitated. While clear and coherent public policies regarding research as well as IP regulation should be further implemented at the national level and the harmonisation of national rules and regulations and the organisation of multi-stakeholder forums should be instituted at the supra-national level. These factors should be addressed by all sides involved in the cooperative relationship.

Large-scale challenges remain, however. The fact that Europe has dominated the creation of funding mechanisms, and has greater access to resources as well as to human and infrastructural research capacity, is generally reflected in the division of roles within specific Africa–Europe research collaborations. In order to redress this imbalance and to achieve greater impact on the global FNSSA challenge—as duly recognised by the African governing institutions such as the AUC and NEPAD, together with African research coordinating platforms on agriculture (e.g. FARA)—capacity development in STI in Africa has to be improved. Possible solutions could, for instance, include the implementation of more joint Africa–Europe doctoral programmes such as the ARPPIS-DAAD Ph.D. scholarships scheme in Kenya, or building upon expert consultations in thematic domains to support multi-disciplinary knowledge sharing, joint priority setting, planning and implementation as exemplified by the

FNSSA partnership. Insofar as this endeavour can and should be addressed within the framework of the Africa–Europe R&I cooperation, efforts should focus more on institutionalising collaborative funding programmes rather than using the unequal and politically biased provisions of development aid.

Outcome Testimonial: Long-lasting partnership through Partnership Platforms (dP-Cirad), Joint International Laboratories (LMI-IRD) or Joint International Research Units (UMI-CNRS): The example of RP-PCP in Zimbabwe. *Compiled by Alexandre Caron (CIRAD) and Priscilla Mugabe (University of Zimbabwe) coordinators of the Partnership Platform.*

In line with the agricultural policy of Zimbabwe, the research platform "Production and Conservation in Partnership" (RP-PCP) aims at contributing to sustainable development, nature conservation and improved rural livelihoods through strengthening national research capacities, multidisciplinary approaches and institutional partnerships. It focuses on protected and neighbouring production areas, with the aim to improve the coexistence of agricultural production and the conservation of natural resources for the benefit of rural communities.

Formally established in 2007, it was renewed in 2010 and 2015 until 2020 following external evaluations. The platform mobilises about 50-to-60 scientists from the Centre de Coopération Internationale en Recherche Agronomique pour le Développement (CIRAD), the National University of Science and Technology, the University of Zimbabwe and the Centre National de la Recherche Scientifique (CNRS). In 2014, it served as the driver of a project called DREAM, which was designed to strengthen and facilitate the links between research and development and its beneficiaries. Given the recognition it received and its high level of achievement, the platform has now entered a phase of institutional and regional expansion.

Under the Africa–Europe FNSSA partnership on R&I, leading researchers in bi-regional collaborations have agreed on the need to sustain their work and ensure observable outcomes are achieved beyond the lifetime of individual projects. However, achieving this is easier said than done. Generally speaking, a number of prerequisites are required, including adherence to the principles of equal representation and collaboration, an in-built element of capacity building, and co-ownership through co-financing and inclusive co-governance. Furthermore, communications and decision making should be transparent.

For most FNSSA projects, participatory approaches involving all stakeholders are essential throughout the project cycle. A participatory approach can also serve as an entry point to better connect the research teams, private sector actors (including farmers' organisations, SMEs, and intermediaries) and decision-makers. In this regard, the projects funded under the ERAfrica framework could serve as a practical template for similar initiatives. Finally, Africa–Europe research platforms dedicated to agricultural research for development (AR4D) and FNSSA should be systematically mapped, their outputs evaluated and their practitioners linked with each other to form a strong community of practice.

Outcome Testimonial: Partnerships to improve irrigation management in small scale agriculture. *Compiled by Jochen Froebrich (Wageningen University, Coordinator of the EAU4FOOD project).*

Increasing agricultural productivity in Africa has long been a pressing issue and a key means to improve the livelihoods of people living in rural areas. Beginning in July 2011, we embarked on a mission to improve irrigation management in small-scale agriculture in Africa through the EU-sponsored project EAU4FOOD, involving several EU-based organisations working in partnership with selected African countries. With a total budget of 4.9 million euros our main objective was to improve agricultural productivity through innovations in irrigation. A key element of the project was the direct involvement of local stakeholders in the design, testing and dissemination of new and more effective soil- and water management strategies. The so-called "Green Wheel Approach" was designed to involve stakeholders ranging from farmers, water managers and retailers to policymakers and non-governmental organisations alike.

Study sites were located in Mozambique, South Africa, Tunisia, Mali and Ethiopia in order to cover every region of the African continent and to obtain a baseline of usable data. In South Africa for instance, two cooperative farms were engaged to collaborate on increasing the yields for tomatoes by acquiring a better understanding of water scarcity in the region and by establishing better links to markets. Another example is the case study of Ethiopia where innovations were tested that aimed at tackling crop pests and improving soil fertility. The results observed in this study led to interest from the government of Ethiopia to further support maintenance of the irrigation infrastructure. Projects in the remaining study sites achieved similar results in terms of improving irrigation and soil fertility and eventually securing higher yields, and thus greater incomes for farmers.

Through this project we have gained direct experience of how inclusion can make a difference. We developed innovations in partnership with local farming communities and involved them in a process to come up with new practices and new ideas for agricultural practices. The EAU4FOOD project inspired new ways of managing irrigation and soil fertility and thus led to an increase in agricultural productivity and minimised the level of pollution of fresh water reserves. Additionally, the project led to changes in agricultural policy processes at national and trans-national levels. As such, it provided an observable contribution to furthering sustainable rural development in Africa by improving the understanding of the importance of irrigation for smallholder farmers.

The Need for Alignment

FNSSA is a complex, multi-dimensional, multi-sectoral issue with links to health, sustainable economic development, environment and trade. STI cooperation can operate in multiple dimensions and via numerous impact pathways. The policy priorities for Sub-Saharan Africa, as stated in the CAADP-FAFS, are to improve the physical and economic access to food and improve utilisation, especially to ensure a diverse diet and increase protein and micro-nutrient supply. Yet, the major focus of research cooperation projects tends to be developing data/knowledge bases, knowledge and information sharing platforms (26%), with only 18% dedicated to food availability and 13% to utilisation. This suggests that current research cooperation projects might be too experimental and not concerned enough with "bread and butter" issues.

The paucity of data on soils and water scarcity, and the need for improving yields, as reported by several leading organisations including the FAO, suggest that future research collaborations for tackling the global FNSSA challenge might do well to target this basic ecological dimension of FNSSA (currently the case for only 12% of projects) (CAAST-Net Plus 2014). Further applied research is needed into the mechanisation aspects (including irrigation) of FNSSA, in particular the role of small and medium-scale energy-efficient equipment and machinery. Since only a small percentage of projects currently appear to focus directly on food access, more attention is also needed on infrastructural development such as farm-market linkages as well as storage and warehousing systems. In addition, along with issues such as risk assessment for minimising the introduction of pathogens into predominantly European food markets, intellectual property rights and bio-based extracts for cosmetics, food safety is important for Sub-Saharan Africa and needs further attention under the FNSSA partnership.

Outcome Testimonial: Innovative organic fertilisers to improve food security. *Compiled by Erick Rajaonary (Chief Executive Officer: GUANOMAD, Madagascar).*

Producing more and better food is vital for securing better lives for millions of people around the world, especially in Sub-Saharan Africa where one in four people remain undernourished. Part of the solution to address this challenge lies in bridging the gap between the private and the public sector. The recurring Africa-Techno Conference, organised by the French organisation CVT, exists to present technologies developed in Africa or Europe that address, among other issues, food security and agri-food. The aim of this event is to identify potential partners or investors in order to scale up the use of innovative solutions to tackle a range of societal challenges. One such technology led to the creation of GUANOMAD in 2006, a Malagasy SME that was supported by the EU-funded African Agriculture SME Fund.

GUANOMAD is a producer of fertiliser drawn from bat excrement. The fertiliser can be used for various crops and is suitable for a broad range of customers. On average 500 kg of GUANOMAD is needed to cultivate one hectare of rice in the first year, while only 425 kg and 380 kg of GUANOMAD fertiliser is needed for the second and third year respectively. Thus, fertiliser use is decreased while harvest yields remain stable and the quality of products improves. This enables the production of organic vegetables, fruits and other horticultural products that in turn help improve food- and nutrition security. GUANOMAD is certified by Ecocert (an organic certification organisation) and thus reduces the use of chemical fertilisers, which benefits the environment.

As part of its funding GUANOMAD also benefitted from technical assistance facilities that included 60.000 euros for export strategy & distribution as well as 250.000 euros for agri-dealer training. As a result, the SME has established contact with agri-dealers in Africa, Europe and the US and are exporting the fertiliser to more than 30 countries. Through the agri-dealer training programme, 100 distributors in the GUANOMAD network benefitted from training to provide technical assistance to local communities and farmers' organisations on how to use the organic fertiliser. The support helped to strengthen the distribution network of GUANOMAD and enabled a scaling up of its operations.

Due to its success of contributing to food security, GUANOMAD has been showcasing its business model at various international conferences. Its involvement in CAAST-Net Plus was a vital stepping-stone towards achieving this by offering a platform for identifying partners and exchanging best practices.

In terms of geographic participation, the main food insecure countries are located in West, Eastern and Central Africa, whereas the majority of the Sub-Saharan African project participants are located in South and East Africa. Western and Central African countries such as the Central African Republic, Chad, the Democratic Republic of Congo and Niger are seldom represented, perhaps reflecting poor STI capacities and networks between these countries and European counterparts. However, some French research agencies (e.g. IRD and CIRAD) are very active in West and Central francophone Africa, while other EU member states have bilateral projects that also target countries in these sub-regions. This provides a starting point from which to build broader collaborative efforts though, as part of this, consideration should be given to utilising funding mechanisms that minimise a "winner takes all" scenario. This will encourage the pursuit of high-quality scientific endeavours based also on insights and capacities from "outliers" within Africa as well as (at both organisational and country levels) to address the global FNSSA challenge.

On the European side, there tends to be low engagement from Eastern European countries even though they face their own related challenges such as food safety and quality, EU market access/penetration, poor infrastructure, and poor policy instruments. As such, these countries could surely benefit from collaborating with African counterparts dealing with similar challenges. Yet, countries such as Bulgaria, Latvia and Poland are poorly represented in Africa–Europe projects, as opposed to Western European countries such as France, Germany, the Netherlands and the United Kingdom. On the African side conflict-prone areas such as Sudan and the Democratic Republic of Congo are hardly covered either; while there are historical causes for this, which can hardly be reversed overnight, this should not discourage ever-increasing efforts to foster the participation of other EU member states, to bring fresh ideas to the fore and to address shared challenges together.

This chapter has thus far focused primarily on the benefits of bi-regional STI collaboration for African FNSSA, saying relatively little about the reciprocal benefits for Europe. To a large extent, this reflects the fact that the relationship between the two continents has a baggage of long-standing inequality, itself the product of vastly different paces in development, notably as regards STI capacity and infrastructure. For many years, European engagement with Africa was primarily in the form of development aid, which, while very useful for the continent's advancement, has created asymmetries. While there is a real desire and drive to establish greater equality in the Europe–Africa partnership (as testified by projects such as ERAfrica), the legacy of historical imbalance slows the pace of transformation. This is particularly true since the capacity building and infrastructure development, required to achieve full equality of means, are still reliant on a greater contribution from the European side even in collaborations based on equal partnership. At the same time, African states could play a "mentoring" role in STI development, which would translate into greater global influence, by pursuing partnerships not only with the larger, historically more familiar European partners but with smaller, less developed European countries (notably in Eastern Europe as noted above) as well as on a regional level.

Having largely addressed the problem of food production in terms of sufficient quantity, Europe is now steadily pivoting towards efforts to increase the nutritional benefits of its yields while also retaining biodiversity, adapting to climate change and reducing greenhouse gas emissions. This effort is exemplified by the Joint Programming Initiative on Agriculture, Food Security and Climate Change (FACCE-JPI), which unites 22 EU countries in addressing "the interconnected challenges of sustainable agriculture, food security and impacts of climate change" (www.faccejpi.com, 2017). Here Africa has many insights to offer. While the primary concern is still overcoming recurrent food shortages, a number of existing research initiatives focus specifically on improving the nutritional quality of food such as Folate Intake in European and African Countries, an ERAfrica-funded project that seeks combat vitamin deficiency by increasing the folate contents of traditional cereal staple foods through fermentation. Or in developing agricultural practices aimed at maintaining biodiversity, exemplified by a collaboration between CIRAD and the University of Pretoria around the cultivation of rooibos.

Research projects designed around the concept of equal partnership as applied not only to input but also benefit would ensure a two-way flow of information and value-addition alike, allowing both Africa and Europe to gain from the collaboration in equal measure. Fortunately, there is a real awareness of this fact: thanks to the Africa-Europe dialogue, it has become clear that the European model for agriculture is being questioned and African policy makers are pursuing models more appropriate to their contexts. Policymakers of both continents are conscious of the fact that we live on a single, interconnected, planet, and that we face the same challenges presented by climate change and the reality of finite natural resources. African and European policy makers have also realised that solutions must be localised and take into account local constraints and specificities. For example, the French Ministry of Agriculture is currently promoting agri-ecology in France based on concepts which were initially developed in Africa, while Europe in turn has launched a programme named LEAP-AGRI (http://www.leap-agri.com, 2017) to support Africa–EU partnerships on priority topics of the FNSSA roadmap (European Commission 2016).

Conclusion

A wider and more intense cooperation is needed in order to strengthen African and European STI policies and practices for greater FNSSA. Such cooperation should strive to draw partners closer together, to harmonise their skills, capacities and resources, while systematically ensuring the equal representation in and ownership of collaborative ventures. Everything else should follow naturally.

Outcome Testimonial: The HLPD and the LEAP-AGRI Project. *Compiled by Johan Viljoen (IRD, Project Manager for CAAST-Net Plus).*

In shared recognition of the importance of STI for societal- and economic growth, the HLPD emerged as the governing body of the JAES STI partnership. Understanding the vital role played by FNSSA as part of the process of development and growth, and in the face of increasing food scarcity and global hunger, the HLPD catalysed the creation of an Africa–Europe Research and Innovation Partnership in FNSSA aimed at proposing both short and long-term actions in order to address shared challenges in this regard. This partnership, in its conception, was to be co-owned and co-funded, as well as aligned with all the major policy developments in both Africa and Europe in the field of FNSSA. The FNSSA partnership is guided by the so-called "FNSSA roadmap", a strategy based on four priority themes meant to serve as basis for a joint Africa–Europe research plan: sustainable intensification, agriculture and food systems for nutrition, expansion and improvement of agricultural trade and markets, and a number of cross-cutting topics.

Within this framework, the LEAP-AGRI project was initiated as flagship collaboration with the objective of increasing joint Africa–Europe investment in FNSSA so as to reduce fragmentation in the field, involving 22 European and African countries and a total budget of more than 22 million euros. In addition to the proposed funding of new joint research projects, LEAP-AGRI seeks to identify and develop existing instruments for cooperation between the two continents, more particularly to involve the participation of the private sector, development organisations and civil society. Guided by the governance principles of partnership, equal participation and long-term commitment, LEAP-AGRI operates within the funding framework of the EU Horizon 2020 programme, and expects among many other results to produce a comprehensive joint Strategic Research and Innovation Agenda for EU–Africa FNSSA.

References

CAAST-Net Plus. (2014). *Africa-EU research collaboration on food security: A critical analysis of the scope, coordination and uptake of findings*. Available from: https://caast-net-plus.org/object/news/1212/attach/CN__FoodSecurityReport_v7.pdf. Accessed 16 May 2017.

CAAST-Net Plus. (2016). *Framework conditions for bi-regional cooperation in the field of food and nutrition security*. Available from: https://caast-net-plus.org/object/document/1626/attach/D1_4_FINAL_WEB.pdf. Accessed 9 May 2017.

European Commission. (2009). *A strategic European framework for international science and technology cooperation*. Available from: http://ec.europa.eu/research/press/2008/pdf/com_2008_588_en.pdf. Accessed 9 May 2017.

European Commission. (2016). *Roadmap towards a jointly funded EU-Africa research & innovation partnership on food and nutrition security and sustainable agriculture*. Available from: https://ec.europa.eu/research/iscp/pdf/policy/eu-africa_roadmap_2016.pdf. Accessed 9 May 2017.

Forum for Agricultural Research in Africa. (2013). *Science agenda for agriculture in Africa (S3A): A report of an expert panel*. Accra: FARA.

NEPAD. (2014). *On the wings of innovation: Science, technology and innovation for Africa 2024 strategy (STISA-2024)*. Available from: http://www.hsrc.ac.za/en/events/seminars/science-tech-and-innovation-strategy. Accessed 16 May 2017.

CHAPTER 5

Africa–Europe Collaborations for Climate Change Research and Innovation: What Difference Have They Made?

James Haselip and Mike Hughes

Abstract This chapter critically assesses Africa–Europe collaborations on climate change research and innovation. Its authors argue that the complexity of research and innovation challenges on this topic calls for subtler collaborative and evaluation programmes. More importantly, they emphasise the need for greater harmonisation between scientific and political priorities on climate change, and point out that project goals should be more precisely defined, so as to ensure that results can be measured concretely and solutions can be progressively improved. In the absence of this clarity, they argue, climate change research and innovation programmes run the risk of being reduced to mere rhetorical statements.

Keywords Innovation • Scientific & political priorities • Economy • Climate Change • Societal challenges • Project goals • Outcome thinking • Outcome mapping • Implementation

J. Haselip (✉)
UNEP DTU Partnership, Copenhagen, Denmark

M. Hughes
Ministry of Education, Kigali, Rwanda

A. Cherry et al. (eds.), *Africa-Europe Research and Innovation Cooperation*, https://doi.org/10.1007/978-3-319-69929-5_5

Introduction

Climate change poses a major operational and strategic risk to economies, ecologies and societies across the world. The specific impacts of climate change, however, are uneven, with some regions and countries experiencing stronger disruptions than others. There are also significant differences in the ability of regions and countries to adapt to climate change: some are already on a strong footing because of their scientific and technological prowess, and others lack basic capacities in research, engineering and policy formulation. In the context of relations between Africa and the European Union (EU), there are fundamental areas of mutual interest when it comes to climate change mitigation and adaptation. These areas are reflected in high-level strategic agreements, such as the Joint Africa–EU Strategy (JAES) (African Union & European Union 2007a), which recognise that research knowledge, and the social and technological innovation it can lead to, has a cross-cutting role to play in addressing the common strategic objectives shared by African and European countries.

In this chapter, we reflect critically on the landscape of Africa–Europe collaboration for climate change research and innovation (R&I). Our guiding question in doing so is a deliberately searching one: what difference have these collaborations made? More specifically, we discuss three key issues: first, the extent to which Africa–Europe research partnerships on climate change have matched up to the stated bi-regional political priorities; second, how and to what extent the outputs of collaborative research have been translated into observable outcomes; and third, whether the research has influenced the direction of policy, business planning or innovation. The discussion in this chapter takes place in the context of a heightened scrutiny over the effectiveness and strategic value of international research spending as well as of development aid. As such, we aim to contribute to a wider debate about how to enhance Africa–Europe research collaboration in terms of the ability to generate and communicate information of relevance to public policymakers and the private sector (European Union 2014).

Scope of Africa–Europe Research Collaboration on Climate Change

What are the joint Africa–Europe research priorities for climate change? The answer to this question is unfortunately not a straightforward one: even though the JAES is the overarching strategy for Africa–Europe cooperation at a bi-regional level, extracting priority topics from the JAES and its action plans is remarkably difficult (African Union & European Union 2007b, 2010). This is especially true for the second JAES action plan (2011–2013): the objectives and expected outcomes are very broad, with an apparent lack of coherence between the priorities stated in the overall objectives, the expected outcomes and the priority actions. The objectives, expected outcomes and priority actions are also, in some cases, closely tied to or presented as concrete projects, which, in turn, add to an unclear presentation.[1] Furthermore, when we survey the scope of the portfolio of Africa–EU collaboration projects funded by the EU Framework Programme (FP), there seems to be an imbalance between the political priorities expressed through the JAES (to the extent that these can be derived) and actual research. The lack of a clear statement of joint priorities on climate change presents a fundamental challenge to the task of assessing bi-regional climate change research projects against the stated political priorities.

An attempt at highlighting some of the priority topics in the JAES can however be made by taking the priority actions listed in the action plans as representative of bi-regional priorities. Using this approach, the relevant topics on climate change emerged as:

- Desertification
- Climate information and earth observation
- Adaptation
- Forests
- The capacity of African negotiators
- Disaster risk reduction
- Biodiversity conservation
- Natural resource management
- Adaptation and mitigation strategies
- Carbon markets
- Climate-friendly technologies

Extracting political priority topics from the action plans does, however, run the risk of excluding topics that are integrated in each priority action. For example, this could be true for a topic like water. Water is not highlighted as a priority in either of the plans. It is however mentioned as forming part of one of the activities in the African Monitoring of the Environment for Sustainable Development project ("Enhancing the African capacities for the operational monitoring of climate change and variability, vegetation, water resources, land degradation, carbon dioxide emissions, etc.") (African Union & European Union 2010, p. 49). While water may be widely viewed as a "big issue", it is mentioned in the JAES in but a single bullet point, for one activity, and under just one priority action. This leads us to conclude that water is *not* a top priority in the JAES.

For our analysis in this chapter, of the topics covered in bi-regional climate change research, and how these relate to the political priorities, we have used the topics listed above as a starting point. To arrive at an overview of Africa–EU bi-regional climate change research projects, we screened 41 relevant FP6 and FP7 projects, and then conducted interviews with managers from 7 projects (see CAAST-Net Plus 2014).[2] The 41 projects we selected were categorised in terms of their primary focus: climate change mitigation, adaptation or both (see Fig. 5.1).

It is evident that there has been more emphasis on adaptation than mitigation or adaptation/mitigation projects in Africa–EU research collaboration on climate change. The split demonstrates a degree of

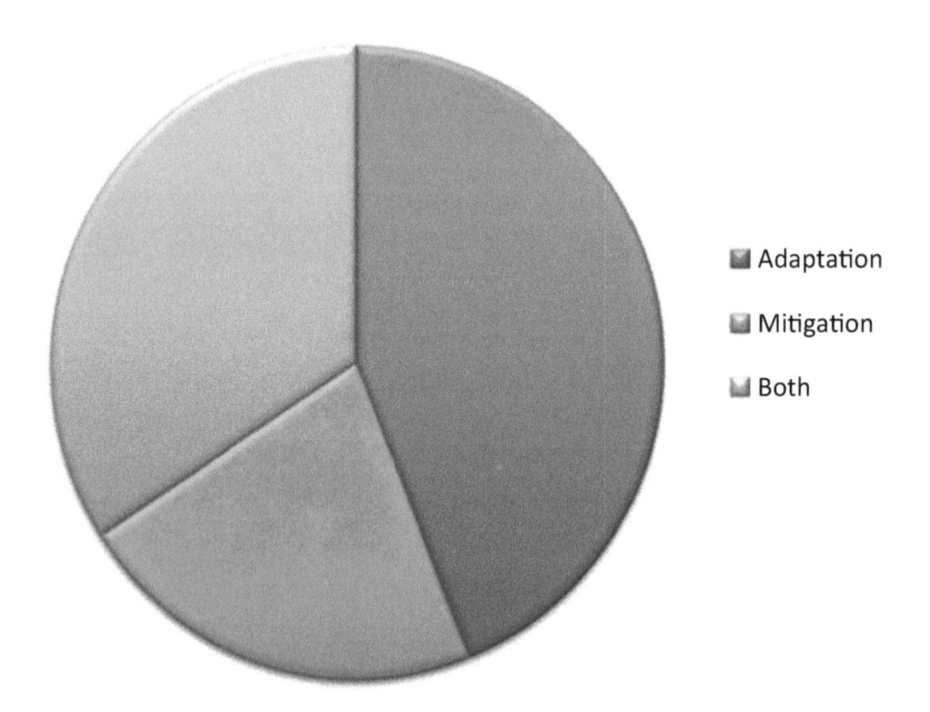

Fig. 5.1 Division of FP6 and FP7 projects according to overall topics (Source: CAAST-Net Plus 2014, p. 5)

coherence between funded projects and the list of bi-regional priority topics, in this case adaptation.

The division between adaptation and mitigation projects, in terms of the size of EU funding, reveals a slightly different picture. On average, mitigation projects received approximately 9.1 million euros per project, while adaptation projects have an average budget of about 6.3 million euros. Projects covering both adaptation and mitigation have even smaller budgets, averaging 5.2 million euros per project. Overall, there is still more FP finance directed towards adaptation than mitigation (Fig. 5.2).

The 41 projects were also divided according to the priority topics given above, some of which cover both mitigation and adaptation issues. In the categorisation of projects in this way, none of the topics are mutually exclusive, meaning that one project can cover several topics. This does not, however, count for the "Other" category, which only includes projects that do not cover any of the other topics:

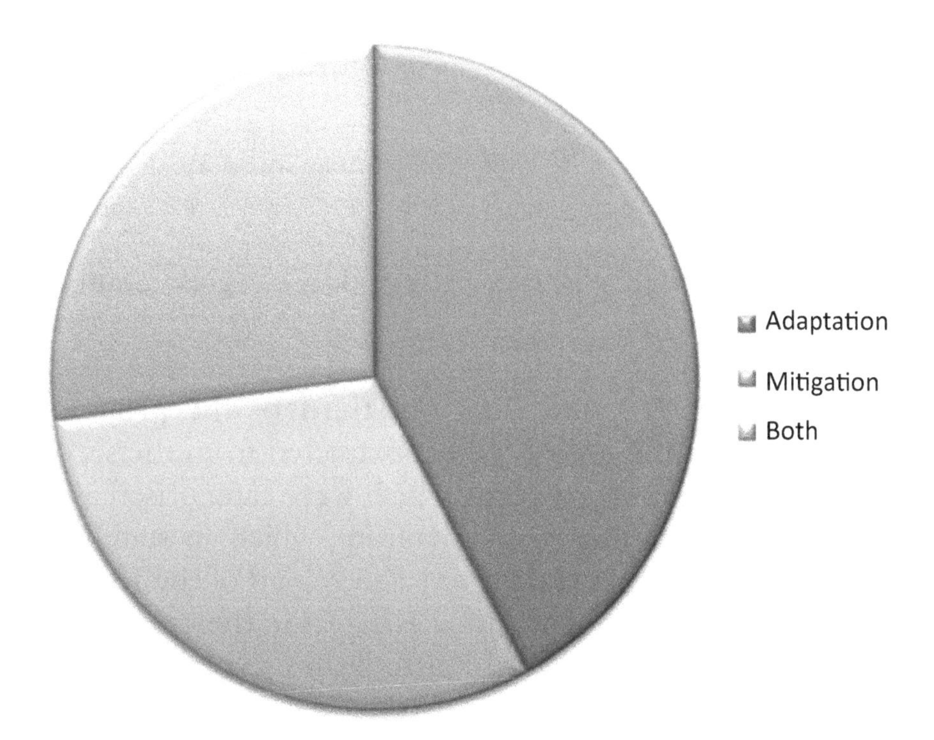

Fig. 5.2 Share of FP6 and FP7 funding spend on adaptation/mitigation (Source: CAAST-Net Plus 2014, p. 5)

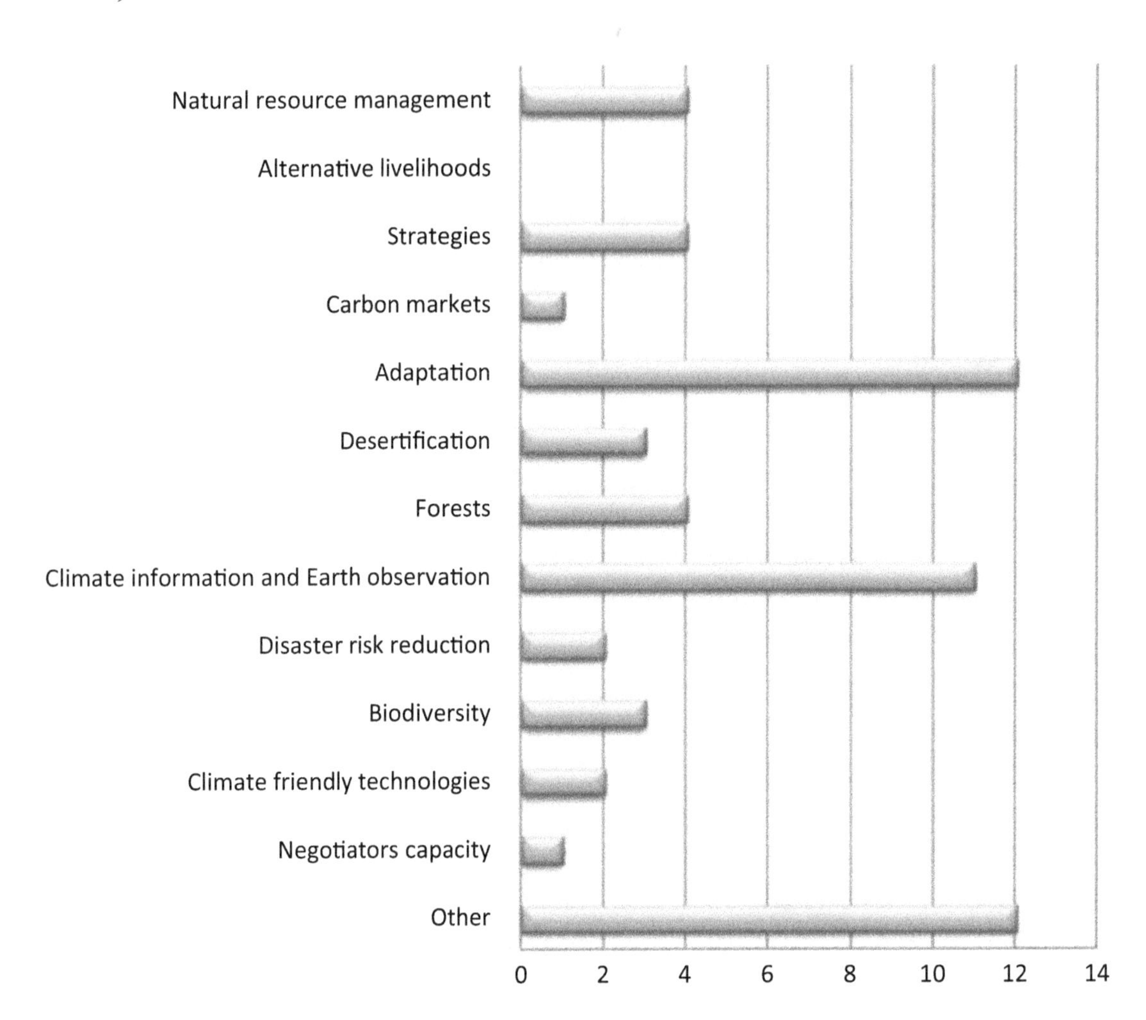

Fig. 5.3 Number of projects in each JAES priority category (Source: CAAST-Net Plus 2014, p. 5)

According to Fig. 5.3, less than one-third of the projects do not explicitly cover any of the priority topics extracted from the second JAES action plan. Several of these projects, which were categorised as "Other", have an explicit focus on water or agriculture, which, as stated above, do not seem to be prioritised in the second action plan of the JAES.

If water and agriculture are included as topics in the categorisation, the distribution looks different. Figure 5.4 indicates that these topics are in fact very prominent in the bi-regional climate change research environment. This is especially true for water, which is included as a focus area in almost half of the projects investigated in the research reflected in this chapter.

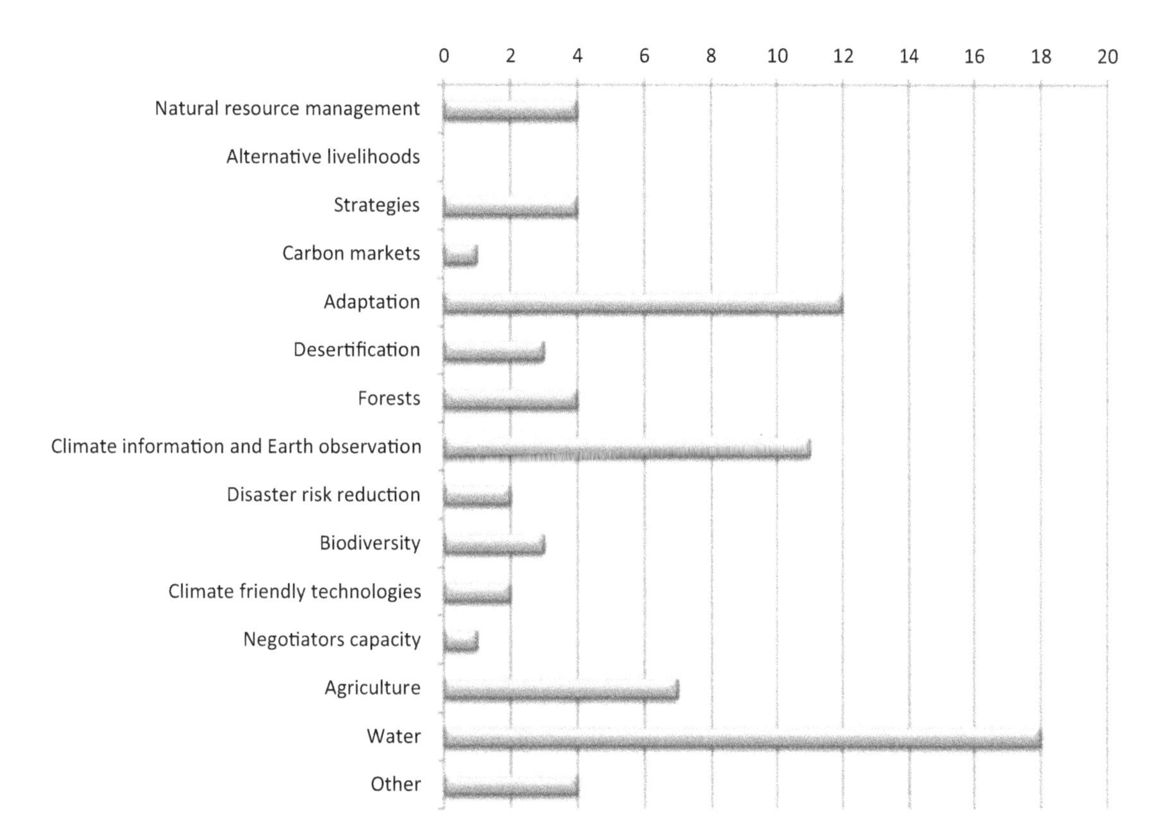

Fig. 5.4 Number of projects in each JAES priority category (including water and agriculture) (Source: CAAST-Net Plus 2014, p. 5)

This thematic focus correlates poorly with the JAES action plans for climate change, which, at best, have a secondary focus on water and agriculture. It is also important to note that a topic such as adaptation covers a wide range of different projects with different thematic emphases, not all of which are listed in the priority topics of the second JAES action plan. As such, there seems to be an imbalance between the political priorities and the actual research conducted, partly explained by the fact that the FP6 predates the JAES. While some of the political priorities are well covered by research, others, like carbon markets, negotiator capacity, disaster risk reduction and climate friendly technologies, are not at all prominent in bi-regional research on climate change.

In drawing these conclusions, it is important to take into account the time lag between the adoption of a political strategy and its manifestation in research projects. This is especially the case for our analysis in this chapter: the JAES action plans cover the period from 2008 to 2013, whereas some of the projects analysed date as far back as 2004. In addition, while we take the list of topics at face value, it is perhaps naïve to assume that there has been any conscious effort on behalf of project developers to interpret or otherwise respond to the JAES priorities. Nonetheless, it is important to know whether there exists a thematic overlap, by accident or design.

Outcomes of Bi-regional Climate Change Research Collaboration

The research-output-outcome chain can be seen as comprising a series of stages, starting with research design and the identification of specific user constituencies and the public at large. Dissemination could be directed at policymakers, and further onwards to various categories of practitioners. The interface with policymakers could lead to policy change or improvement. In turn, policy change or influence could lead to uptake by practitioners and users. Only the end result of these convoluted processes could qualify for the term "impact". Owing to their complexity, outcomes are usually better represented through narratives. Generic indicators or quantitative measures can only count outputs that in themselves are of little meaning in terms of pointing to the ultimate impacts of a particular initiative. In this sense, we draw in this chapter on the so-called outcome mapping school of thought (see www.outcomemapping.ca, 2017).

In order to investigate the difference that any given project or intervention has made, it is useful to first document the *intended impacts*, as conceived and pursued by project managers. When research projects are granted funding under the FP, for example, project descriptions usually contain statements of expected "impact" that is used as a criterion of project assessment. We asked respondents to describe the expected impacts of their projects, as defined at the start of the project. They were also asked if and how they tried to measure impacts and, the means or methods employed. Furthermore, we asked if they were able to plausibly attribute the observed "impact" to their specific research project, rather than to other intervening factors. We wanted explanations of *how* these outcomes occurred, that is, we asked: What was the "mechanism of change" at work during and after project implementation? With this type of investigation outcomes can only be linked to a specific activities through plausible claims (i.e. reasonable arguments provided by stakeholders as to the cause-and-effect relationship between the identified research project and a given policy, practice or behaviour).

In the case of the African Monsoon Multidisciplinary Analysis (AMMA) project, the following bullet points are the closest to a statement of intended impacts:

- To assist in the achievement of the United Nations (UN) Millennium Development Goals (MDGs) in Africa and the implementation of the EU Strategy for Africa, which includes "action to counter the effects of climate change" and "the development of local capabilities to generate reliable information on the location, condition and evolution of environmental resources, food availability and crisis situations"
- Add to the African participation and ownership of AMMA research activities, and strengthen the linkages between European research institutions and the West African research community
- Ensure that the further development of national expertise is maintained beyond the AMMA project

While such statements sound plausible and convincing, they serve mostly to highlight the topical relevance of the research. Indeed, according to Jan Polcher, European leader of the AMMA project, "the impact section of the proposal was very much political talk" (CAAST-Net Plus 2014, p. 32). Similarly, the major anticipated outcomes of the FP7 project ClimAfrica focused on:

- Responding to environmental degradation as relevant for poverty alleviation and food security enhancement
- Specific climate change mitigation and adaptation options for local communities
- Capacity of team members and other stakeholders within the communities enhanced
- Synergies with existing actors (NGOs, district assembly etc.) in the various localities strengthened

These typically vague statements of intended impact are difficult to measure, or verify. Ernest Ohene Asare of the Department of Physics at the Kwame Nkrumah University of Science and Technology (KNUST) in Ghana, which is also a "beneficiary" of the AMMA and QWeCI

(Quantifying Weather and Climate Impacts on Health in Developing Countries) projects, offered a more concrete account of observation outcomes. Project funds, Ohene Asare said, were invested in the acquisition of instruments needed for data collection and therefore better data were collected for the AMMA and QWeCI projects (CAAST-Net Plus 2014, p. 32). Specifically, he worked on a malarial model to be used in Ghana with partners from University of Cologne, Germany, and the International Centre for Theoretical Physics, Italy, and is currently working to improve understanding of the breeding temperature of mosquitoes with the help of colleagues at KNUST.

Ohene Asare stated that the project helped him to "get exposure" and that he was able to work with other scientists, which also enhanced project work and gave new directions. He added that his presentation skills improved and that he learned how to communicate and disseminate the results of scholarly work through tailor-made presentations, personal discussions and formal interviews. More importantly, the two projects have brought together scientists from a range of disciplines and have fostered networking and knowledge sharing. While this account is more concrete, it nevertheless falls short of responding to the project's statements of intended impacts, and rather provides an anecdotal basis for attributing project outcomes.

The AFROMAISON project makes reference to "impact pathways", developed at the beginning of the project. In this approach, potential impacts are identified, elaborating the mechanism of change in a participatory manner by involving key stakeholders from the outset. As the project managers noted, this serves as a guide to implementation, a means for periodically checking whether the "impact theory" is correct, and making adjustments during implementation. If properly followed, this appears to be an effective approach for enabling the identification of realistic outcomes and how the project outputs can lead to these. Generally speaking, the articulation of impact pathways is considered to be part of the challenge of enhancing the ownership of tools and empowerment of the subnational authorities and communities.

In Uganda, the AFROMAISON project developed scenarios to understand the extent to which human activities had an impact on the natural resources and ultimately on the climate in the Rwenzori mountains/ Albertine region. The approach taken was through "action research", where the project team and communities met to share experiences and agree on practical solutions for pertinent natural resource management (NRM) challenges like landslides, silting of rivers and floods. These scenarios served as both awareness raising tools and consensus building platforms for effective NRM. They also helped to ensure that research results were acceptable and directly beneficial to the target community. A key project output was the development of a participatory tool, "Mpang'ame", a simulation game that helps stakeholders identify and reflect on appropriate actions for better NRM practices. At the local level, the game was disseminated to schools, vocational institutions, local government leaders and policymakers within various fora. At regional and international levels, the game was disseminated at meetings for AFROMAISON partners and NRM stakeholders in Burkina Faso, Ethiopia, Mali, South Africa, Tunisia, as well as to graduate students in France, on special request from university administrators. Other dissemination channels included articles published in the *International Journal of Innovation Sciences*, book chapters, brochures, leaflets and the project website.

According to Arseni Semana, principal investigator of AFROMAISON in Uganda, the main challenges that the project encountered were related to the attitudes of the communities (CAAST-Net Plus 2014, p. 32). There was slow adoption of integrated NRM practices mainly because of the commercial culture that has emerged within the beneficiary communities. NGOs facilitate communities' participation in NRM planning and implementation. As a result, it is almost impossible to engage the communities without attaching a monetary incentive. Private sector involvement is still minimal and participating private sector players are mainly informal and micro. Nonetheless, the project held a consultative meeting between the ministries of agriculture, animal industry and fisheries, and water and environment to enhance policy level integration of NRM using tools from the research. This constitutes a more valuable, critically reflective account of the relationship between the project's outputs and outcomes, and one which integrates key contextual factors to explain the barriers and constraints to achieving the intended impacts.

Overall, we a found that statements of "intended impact" are often more akin to aspirations expressed by project designers and managers. In most cases, these aspirations do little more than offer rhetorical support to wider climate and development targets, such as the MDGs. As such, there is generally no explicit explanation of *how* these impacts can—even theoretically—be achieved. Instead, there is a significant level of assumed attribution; that is, broad statements about how the research project's focus relates to the wider issues and how it contributes knowledge necessary to tackle these challenges vis-à-vis the stated aims and objectives.

We also found that very few FP projects make clear distinctions between outputs, outcomes and impacts at the design stage. Consequently, the terms are often confused or used interchangeably. The most common mistake is to present and refer to project *outputs* (workshops, research articles, policy papers, conferences etc.) as *outcomes*. Similarly, there is an over-referencing by project designers who, in outlining their activities, indicate "engagement with a variety of stakeholders" as key. This is simply another rhetorical device that, while politically correct and plausible, is rarely explained in detail and hence fails to substantiate a convincing theory, or set of mechanisms, for actual change.

While our interview schedule placed a sharp emphasis on understanding how outcomes and impacts were understood and anticipated, our questions often proved difficult for respondents to answer. The latter often drifted towards a focus on more procedural and "mundane" aspects of Africa–EU research collaborations, including the challenges of day-to-day management and the ultimate delivery of project outputs. Or they focused on the challenges of coordination and of targeting key project conclusions or recommendations to the most appropriate audiences. If they did manage to engage with "target audiences", then there was often little or no follow-up that would enable project managers to understand the extent to which these key messages had influenced policymakers or the business community. Information and knowledge in this regard remain anecdotal, at best. It was also a challenge to receive concrete examples of "outcomes", as understood in outcome mapping analysis, which many respondents confused with "outputs". This is a fundamental issue and one that appears to explain the paucity of plausible arguments to attribute project outputs to demonstrable outcomes.

Engaging and Influencing Public and Private Decision Makers in Africa and the EU

To what extent have the research and development outputs from Africa–EU climate change collaboration, funded by FP6 and FP7, informed public policymaking and business planning? It is widely acknowledged that applying technical knowledge to policy and business planning is a key challenge. But what do we know about the barriers and constraints to such uptake? How can these be removed? We attempted to answer these questions by analysing responses from government, civil society and commercial actors. Our concerns centred on the issues faced in specific projects, such as: What were the main challenges in communicating research to a policy and business audience? Did project partners interact with policymakers? If so, did these actors adopt the research findings as evidence in support of their policy formulation or revision, and how did they ascertain whether they did so or not? If they did not embrace the research findings, what was the reason?

Principal investigator of the AMMA and QWeCI projects in Ghana, Sylvester K. Danuor of the physics department of KNUST, said that in order to achieve the project's intended impacts, research findings were disseminated mainly at conferences and workshops, and through journal articles. According to Danuor, workshops were the most effective means of reaching out to the intended beneficiaries. These included the research community, policymakers and civil society organisations. He and other interviewees were of the view that the AMMA and QWeCI projects "had some interaction" with policymakers who "embraced the research findings" (CAAST-Net Plus 2014, p. 32). However, this was yet to be reflected in official policy formulations. For instance, there were meetings with the Ghana Meteorological Agency and the District Health Directorates through the Metropolitan Health Directorate of the Ministry of Health. There was a similar positive interaction with civil society organisations with a view to encouraging them to make use of the project's findings in policy formulation and activities.

This account of project–policy interactions is typical of the responses we received. These responses reveal a high degree of uncertainty and inability to verify the claims, however plausible they appear. As already mentioned, this reflects a lack of "outcome thinking" at the level of research project design and management. In short, there was a predominant focus by project managers on outputs that are easy to document and report. Where an interaction with policymakers is mentioned, the precise mechanism through which research outputs actually influence policy or practice is rarely explained in any detail. As such, efforts to engage with and influence policymakers are mostly *ad hoc* at best, and amount to little more than a hope or expectation that the research findings will be accessed, understood and taken up by the relevant actors in government or the private sector. In turn, the lack of clear *mechanisms or theories of change* undermines efforts to reflect on the project implementation process or face the hard question of what difference their efforts made. Finally, there is also a general lack of follow-up studies to monitor longer-term outcomes of framework research projects, which once again reflects the predominant focus on monitoring, reporting and evaluating the strength of project outputs.

In the realm of private sector engagement, there is minimal evidence of FP6 and FP7 research projects generating climate change knowledge that feeds directly into technology development or patents.[3] We would, however, expect to gather at least some anecdotal evidence of positive relationships between research projects and technology developers and/or private sector investors operating in the market for clean and low-carbon technology. To a large extent this lack of obvious examples reflects the thematic focus of many FP-funded projects on climate change: the majority focus on the generation of basic research knowledge, such as emissions monitoring and data analysis, or capacity building, which does not have a strong or obvious commercial application. As such, there are generally low levels of private sector involvement in Africa–Europe research collaborations on climate change, which, by extension, appears to suggest that the FP has had limited success in supporting innovation.

While it may not be easy to identify a clear attribution between Africa–EU research collaborations on private sector innovation and technology development, it does not mean that it does not occur. Indeed, it is far more likely that private sector actors will be drawing on the findings of such research collaborations in the preparation of their business plans, given they have a clear and strong incentive to develop their business and investment intention upon scientifically sound findings. The fact that most FP-funded research findings are publically available would make this even more likely, though the project managers and partners would be unaware of this information uptake.

Another issue that may constrain the active promotion of FP-funded research findings into public and private (non-research) forums is the lack of ability or willingness by project managers to actively engage with such decision makers. In the case of climate change research all the recent framework projects are managed by European-based institutes. This fact may be of material consequence in terms of their limited contact—that is, apart from via project partners—with local policymaking and business leader networks. There may also be reluctance on the part of Europeans to get involved with local policymaking and politics. Project managers are likely to be unfamiliar with the complex institutional and policy terrain of

African countries. As Jan Polcher, the European-based manager of the AMMA project, observed:

> our main targets were the local scientific community and the operational agencies [...].[However] it is my belief that Western scientists have no role in disseminating to policymakers; civil society organisations; politicians; private sector in West Africa. Because of the colonial heritage our message would not have the desired impact. So this dissemination is to be left to the regional research community. (CAAST-Net Plus 2014, p. 32)

This is an unusually frank but significant admission by a project manager who would in principle be responsible for pushing the research-to-policy connections. It raises more questions about whether the research-to-policy agenda is being advanced in the first place, despite the broad statements of intended impact mentioned in the project documents.

Many of the respondents in this aspect of our research focused on the difficulty of directing conclusions or recommendations at most appropriate audiences. If and when they did manage to engage with target audiences, then there was often little or no follow-up that would enable project managers to understand the extent to which these key messages had influenced policymakers or the business community. Knowledge in this regard remains anecdotal, at best.

Conclusion

Even though the JAES is supposed to be the overarching strategy for Africa–Europe cooperation at bi-regional level, extracting specific climate change research priority topics from past JAES action plans proved difficult. The plans should not therefore necessarily be seen as the guiding document for bi-regional research on climate change. What this chapter also showed is that the theories of change inherent in most FP-funded projects—to the extent they are made explicit—are too simplistic and depend upon linear concepts, as manifested in the predominant logframe approach to project design and management. There appears to be a low level of outcome thinking to the extent that many respondents confused project outcomes with project outputs. This is a fundamental issue for Africa–EU research collaborations across thematic areas, and one that appears to explain the paucity of plausible arguments to attribute project outputs to demonstrable outcomes in the context of the collaboration on climate change studied in this chapter.

Furthermore, this chapter argues that statements of intended impact are often tantamount to mere aspirations expressed by project designers and managers, which in most cases do little more than offer rhetorical support to wider climate and development targets, such as the former MDGs. As such there is generally no explicit explanation of how these impacts can, even theoretically, be achieved. Instead, there is a significant level of assumed attribution, that is, broad statements of how the research project's focus relates to the wider issues and how it contributes knowledge necessary to tackle these challenges *vis-à-vis* the projects' aims and objectives. Similarly, there is too much reference to projects aiming to achieve their stated aims and impacts by engaging with a variety of stakeholders, another rhetorical device that is at once politically correct and plausible, though rarely explained in detail and hence fails to provide a convincing theory, or mechanism, of change. Such a lack undermines efforts to reflect upon the project implementation process and to answer the question "what difference did it make?"

Notes

1. Cases in point include the Great Green Wall of the Sahara and Sahel Initiative, ClimDev, African Monitoring of the Environment for Sustainable Development and the Global Climate Change Alliance.
2. The projects included AFROMAISON, AMMA, Animal Change, ClimAfrica, DEWFORA, Healthy Futures and QWECI.
3. It should be acknowledged that this finding is based on an in-depth questioning of a small sample of projects, so caution should be taken in drawing programme-wide conclusions.

References

African Union & European Union. (2007a). *The Africa-EU strategic partnership: A joint Africa-EU strategy*. Available from: http://www.africa-eu-partnership.org/sites/default/files/documents/eas2007_joint_strategy_en.pdf. Accessed 8 May 2017.

African Union & European Union. (2007b). *First action plan (2008–2010) for the implementation of the Africa-EU strategic partnership*. Available from: http://www.africa-eu-partnership.org/sites/default/files/documents/jaes_action_plan_2008-2010.pdf. Accessed 8 May 2017.

African Union & European Union. (2010). *Joint Africa-EU strategy: Action plan 2011–2013*. Available from: http://www.africa-eu-partnership.org/sites/default/files/documents/03-JAES_action_plan_en.pdf. Accessed 8 May 2017.

CAAST-Net Plus. (2014). *Africa-EU research collaboration on climate change: A critical analysis of the scope, coordination and uptake of findings*. Cape Town: Research Africa.

European Union. (2014). *Mapping of best practice regional and multi-country cooperative STI initiatives between Africa and Europe: Identification of financial mechanisms 2008–2012*. Luxembourg: Publications Office of the European Union.

CHAPTER 6

Equality in Health Research Cooperation Between Africa and Europe: The Potential of the Research Fairness Initiative

Lauranne Botti, Carel IJsselmuiden, Katharina Kuss, Eric Mwangi, and Isabella E. Wagner

Abstract This chapter investigates the strategic benefits of global health collaboration programmes. Regretting the lack of alignment or harmonisation of research priorities and cooperation patterns, authors show how recent positive research development on health issues in Africa can foster more constructive and more balanced research partnerships with European countries and institutions. In this regard, authors urge greater support for

L. Botti (✉) • C. IJsselmuiden
Council on Health Research for Development (COHRED), Geneva, Switzerland

K. Kuss
Spanish Foundation for International Cooperation, Health and Social Affairs, Madrid, Spain

E. Mwangi
Ministry of Education, Science and Technology, Nairobi, Kenya

I.E. Wagner
Centre for Social Innovation, Vienna, Austria

A. Cherry et al. (eds.), *Africa-Europe Research and Innovation Cooperation*, https://doi.org/10.1007/978-3-319-69929-5_6

the Research Fairness Initiative as a promising emerging global standard for fostering fair and sustainable research partnerships and a more inclusive and better institutionalised framework for Africa–Europe cooperation on health development and innovation.

Keywords Global health cooperation • Global standards • Sustainable partnerships • Health development & innovation • Ebola • HIV • Malaria • Clinical trials • Collaboration outputs • Quality & fairness • Private sector

Introduction

In recent decades, governments have increased their collaborations on strategies for global health, and multilateral research programmes have involved partners from high-, middle- and low-income countries. Cooperation on health issues between Africa and Europe reveals the need to address the asymmetries that can affect both global health and health research. The outbreak of Ebola in West Africa in 2014 resulted in over 11,000 recorded deaths. With the disease also threatening Europe and rapidly becoming a global issue, it reminded us of the borderless vulnerability of our populations and of our responsibility to invest in global health and health research. Indeed, the Ebola epidemic did influence the international agenda for global health. The European Union Council, for example, stressed the importance of health security in the European Union (EU) and the need to strengthen preparedness research to address health security. Following a renewed interest in global health, the European Parliament also requested the evaluation of the impact of EU Framework Programmes (FP) funding of research into poverty-related and neglected diseases (PRND) on universal health coverage (UHC) (see RAND 2017).

This chapter elaborates the results of a study conducted by the CAAST-Net Plus project, concerning the impact of Africa–Europe health research cooperation under the European and Developing Countries Clinical Trials Partnership (EDCTP), and under the EU FP and its contributions to the broader bi-regional science, technology and innovation (STI) partnership (see also CAAST-Net Plus 2016). The study examined health research cooperation between Africa and Europe and the impact it has on participating countries. The first three parts of this chapter respond to a set of concerns about the extent to which (1) actual bi-regional collaboration matches up to joint research and innovation (R&I) priorities; (2) bi-regional collaboration is balanced; and (3) the outputs of bi-regional collaboration are translated into new or revised goods, services, technologies or new or revised policy. The fourth part of this chapter presents the Research Fairness Initiative (RFI) as a response to the widely acknowledged need for improved quality and fairness in Africa–Europe research collaborations.

Policy Frameworks and Priorities

The main policy framework that currently guides research cooperation between Africa and Europe at regional level is the Joint Africa–EU Strategy (JAES) adopted in 2007 (African Union & European Union 2007) by the member states of the African Union (AU) and the EU at the second Africa–EU Summit in Lisbon. Although science is no longer an explicit chapter of the current JAES action plan, the contribution of STI remains embedded in it. The JAES states unequivocally that health research should address global challenges and common concerns related to HIV/AIDS, malaria, tuberculosis (TB) and other pandemics (paragraph 8), while research on vaccines and medicines for major, neglected and water-borne diseases should be supported (paragraph 61) and national health systems strengthened through the development of integrated strategies (paragraph 61).

The JAES stands out as one of the few frameworks that explicitly outline joint priorities for bi-regional cooperation in health research, although many national and international policies, declarations, strategies and agreements do provide guidelines for policymakers to formulate research cooperation priorities. For example, the Sustainable Development Goals are one of the most influential international agreements that guide and feed into bi-regional cooperation strategy and priorities in health research. These goals directly impacted international strategies and programmes such as the Special Programme for Research and Training in Tropical Diseases hosted at the World Health Organization (WHO), and have led to ambitious initiatives such as the Global Fund to Fight AIDS, Tuberculosis and Malaria and Global Vaccine Alliance. It is important to note that the key issues of access to UHC and to vaccines were addressed by the recent declarations such as the 2014 Luanda Commitment on

Universal Health Coverage in Africa, and the 2016 Addis Declaration on Immunisation (WHO 2014; Ministerial Conference on Immunization in Africa 2017), though both were left out of both the joint Africa–Europe agenda for science and technology cooperation and its implementation roadmaps.

A global analysis of the deaths by infectious diseases and non-communicable diseases (NCDs, such as cancer, diabetes or mental health) concludes that there was an increase in HIV/AIDS, malaria and TB deaths between 1990 and 2010.[1] Mortality due to HIV/AIDS reached a peak of 1.7 million in 2006; malaria mortality rose to 1.17 million deaths in 2010 and TB killed 1.2 million people in 2010. In parallel, NCDs rose by just under 8 million between 1990 and 2010, explaining a third of overall mortality worldwide by 2010 (34.5 million) (Lozano et al. 2012). The numbers of deaths caused by NCDs are clearly increasing rapidly. A report by WHO (2017) has aptly summarised the global burden of NCDs as follows:

> NCDs kill 40 million people each year, equivalent to 70% of all deaths globally. Each year, 15 million people die from a NCD between the ages of 30 and 69 years; over 80% of these "premature" deaths occur in low- and middle-income countries. Cardiovascular diseases account for most NCD deaths, or 17.7 million people annually, followed by cancers (8.8 million), respiratory diseases (3.9 million), and diabetes (1.6 million). These 4 groups of diseases account for over 80% of all premature NCD deaths. (WHO 2017)

Bi-regional health research collaboration matches joint priorities particularly on HIV, malaria and TB. Many African countries have built substantial research capacities on these three major diseases. In 2013 the World Health Assembly adopted a resolution that calls for increased investments to improve the health and the social well-being of affected populations (World Health Assembly resolution 66.12). Almost at the same time, research on neglected diseases (NDs) was included in the second EDCTP Programme—NDs are NCDs that prevail mainly in subtropical conditions and largely affect populations living in close contact with infectious vectors and domestic animals.

Africa–EU health research cooperation does address global challenges and common concerns in terms of malaria, TB and, more recently, in terms of NDs. Nevertheless, health research priorities, as mentioned in the JAES, need to be updated to reflect the changing needs and evolving burden of diseases. In the next 10 to 20 years, estimates predict a dramatic increase in the prevalence of NCDs, which will account for nearly 40% of disease burden in Sub-Saharan Africa by 2030 (Olesen and Parker 2012). This already has consequences for the current R&I collaboration—not just in 10 or 20 years from now. In this context, Africa–EU collaboration will require additional research investments to prevent NCDs through new vaccines, diagnostics and treatment, and to improve and increase access to health facilities and health coverage.

To better understand Africa–Europe science cooperation patterns in health research, CAAST-Net Plus conducted a bibliometric study on health co-publications between Sub-Saharan African and European researchers in recent years.[2] Bibliometric assessments of joint research in health have already been conducted, for example by Breugelmans et al. (2014, 2015) who compared research publications on PRNDs. Both of these studies found an overall increase in the volume of collaborative research outputs, similar patterns in geographic differences and an overall emphasis on PRDs. However, there have been no comparative analyses of the current research areas in Africa–Europe collaboration.

The study conducted by CAAST-Net Plus analyses the volume of publications on HIV, malaria and TB, collectively here called poverty-related diseases (PRDs), as well as on NDs and on NCDs in bilateral cooperation between Europe and Sub-Saharan Africa (SSA).[3] The data was analysed according to the three health research specialisations, defined by keyword sets.[4]

Figure 6.1 shows the development of the three strands of health research specialisations over the last decade. While the overall number of EU–SSA co-publications in health increased steadily (from slightly more than 2000 in 2005 to almost 5500 in 2014), the relative proportion of publications on NCDs, PRDs and NDs changed: publications on ND and NCD grew while fewer publications on PRD were published, although they still constitute the strongest research strand in comparison to the other two.

The increased attention given to NCDs is all the more positive as they have long been ignored although their burden might soon be higher than that of infection diseases. Yet, NCDs are still not a priority, as the number of publications on PRDs has been growing much faster than on NCDs. In fact, African research institutions do not participate fully enough in research on NCDs, as in PRDs or NDs. Several calls to fund research on NCDs have been recently issued by African institutions; for example, the

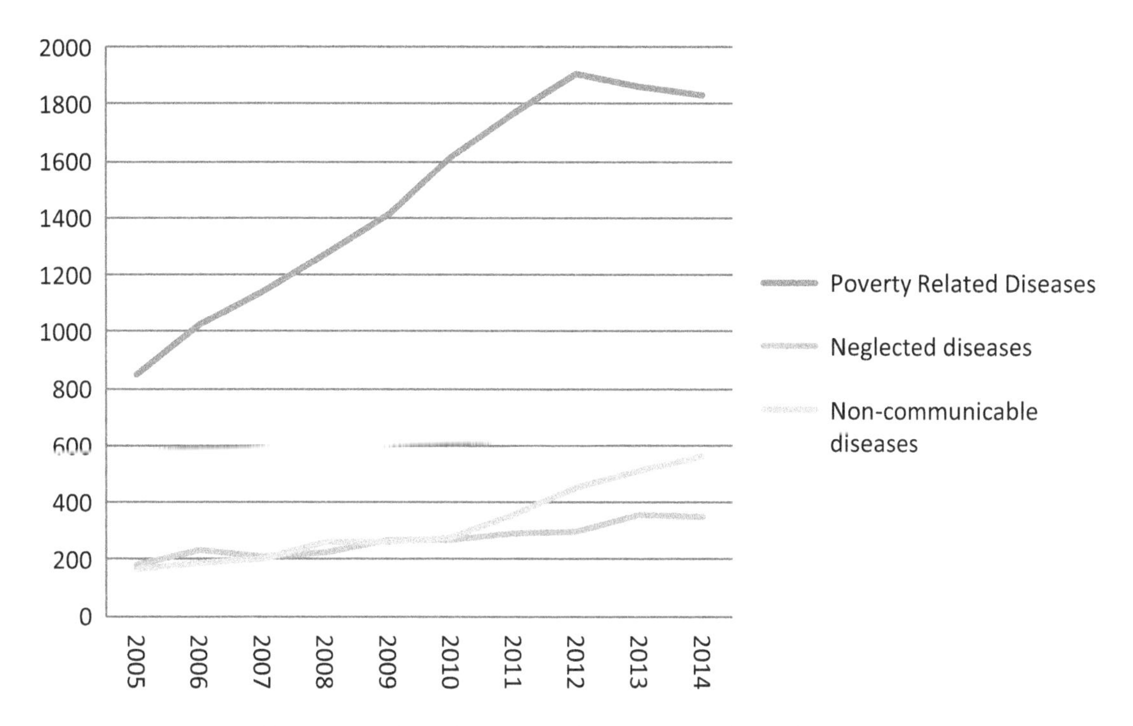

Fig. 6.1 EU–SSA co-publications 2005–2014 in the selected strands of health research (Source: CAAST-Net Plus 2016)

South African Medical Research Council partnering with the Newton Fund and GlaxoSmithKline issued two calls to address the WHO objective to decrease preventable mortality by 25% from NCDs (London School of Hygiene and Tropical Medicine 2015).

Among the pioneers of African research institutions participating in research on NCDs are consortium partners who participated in projects responding to the first FP call for proposals addressing infectious agents and cancer in Africa (HEALTH.2010.2.4.1) and to the second call HCO-05-2014, *Global Alliance for Chronic Diseases: Prevention and treatment of type 2 diabetes*. Three research projects funded by the FP on NCDs have at least one partner from Africa:

1. Prevention of liver fibrosis and cancer in Africa (PROLIFICA): Focusing on women's health, specifically the prevention of cervical cancer by early detection or by vaccination (MRC Unit the Gambia 2017)
2. Human papilloma virus in Africa research partnership (HARP): Evaluation and impact of screening and treatment approaches for

the prevention of cervical cancer in HIV-positive women in Burkina Faso and South Africa (CORDIS 2017)

3. Self-management approach and reciprocal learning for the prevention and management of type-2-diabetes (SMART2D): The project is a member of the Global Alliance for Chronic Diseases and contributes to the Alliance through the development of the community management strategies for the low-, middle- and high-income settings (Karolinska Institutet 2017)

Although the burden of infectious diseases is similar to the socioeconomic impact of those pandemics, many African countries have built substantial research on research in HIV/AIDS, malaria and TB. Over the long term, research dedicated to NCDs could show positive results that would reduce costs for often lengthy and expensive treatment of cardiovascular diseases, cancers, diabetes or chronic lung diseases, and so could contribute to alleviating the socioeconomic burden of NCDs. The accessibility and affordability of healthcare services and products are also major challenges to be tackled, and so are preventive health services. Ideally, the contribution of research projects to health care, health system services and shaping national R&I systems in low- and middle-income countries should be made an explicit objective of all Africa–Europe cooperative research calls.

Working Towards More Balanced Bi-regional Collaboration

Investments in research on PRDs on the one hand, and the increasing burden of NCDs on the other, remain disproportionate. The CAAST-Net Plus study of joint co-publications by authors affiliated to institutions in Europe and SSA shows an increase in publications on NCDs in the period 2004–2015 while the total volume of co-publications remains relatively low. A similar picture results from analysis of research projects funded by the different FPs within the health societal challenge area. Such observations call into question the balance (regarding the scientific and geographical scope, the funds, as well as the ownership and leadership over cooperative project) within bi-regional cooperation.

Nevertheless, EDCTP is a remarkable example of balanced cooperation in terms of governance and participation. Legally, EDCTP is an association established under Dutch law in the Netherlands, which currently counts 28 partner states as full and equal members—14 African and 14 European. Focusing on the development of indispensable research infrastructure, EDCTP has been contributing substantially to the Africa–Europe partnerships, because of its focus on the development of new and improved drugs, vaccines, microbicides and diagnostics against HIV/AIDS, TB, malaria and neglected infectious diseases. Among the results achieved by the programme are (1) the Kesho Bora Study, which demonstrated a 43% reduction in HIV infections in infants and more than 50% reduction of mother to child transmission during breastfeeding and influenced WHO 2010 guidelines on prevention of mother-to-child transmission of HIV, and (2) the Malaria Vectored Vaccine Consortium which found that the volunteers receiving the T cell-inducing vaccine had a 67% reduction in the risk of malaria infection during eight weeks of follow-up (see also EDCTP 2017).

The EDCTP programme, with its comparatively large funding for African institutions, has also become a success story from the perspective of balanced funding. The first phase of the EDCTP lasted from 2003 to 2013 and in this programme, 70% of funding went to African institutions and 62% of all projects were led by African researchers. A significant portion of the funding was aimed at capacity building and support for the ethical and regulatory environment for clinical research in Africa that includes, for example, the African Vaccine Regulatory Forum (the network of ethics committees), National Ethics Committees (NECs), the Mapping of Research Ethics Committees in Africa and the Pan-African Clinical Trials Registry (PACTR).

Critical voices on clinical trials question the balance of benefits for research on the one hand, and the benefits to participants in research on the other. The involvement of patients and volunteers in clinical trials, particularly in low- and middle-income countries, requires researchers to adhere to international guidelines for ethical conduct in health research. The guidelines demand that researchers assess and weigh the burdens to the individuals and groups involved in the research with foreseeable benefits to them and to other groups. Participants in clinical trials often wish or expect to obtain better access to healthcare and products, additional diagnostics tests and treatment or collateral health services that is normally not available. Benefits to populations during research often include ancillary health services such as distribution of medicines or distribution of vaccines. Such criticisms point at important questions that remain to be answered: How can we ensure that health research contributes to better health care? Is there a legal or moral obligation to provide training to researchers and healthcare staff? What about technology transfer and medical equipment?

Depending on the nature, risks and burdens of the collaborative research, mutual negotiations should culminate in agreements or memoranda of understanding (MoU) aimed at providing a fair level of benefits to the host country, research institutions and communities. All clinical trials should be performed in compliance with local ethical and regulatory requirements. Nevertheless, research ethics committees cannot be made solely responsible for preventing unethical or exploitative conduct. Lack of staff, time and resources for follow-up restrain the agency of such research ethics committees. Access to adequate research infrastructure and equipment is critical for the quality of research too—in 2016, South Africa launched the Research Infrastructure Roadmap to improve researcher's access to world-class scientific knowledge and facilitate long-term planning to establish competitive national system of innovation (SAnews 2017). While funders should invest more in targeted equipment and infrastructure grants for African institutions to become internationally more competitive (Doloro 2016), mechanisms going well beyond the review of individual studies are necessary to ensure that partnerships result in systematic national capacity building in R&I. As the last section of this chapter explains, the RFI was precisely designed to go "beyond ethics review".

The CAAST-Net Plus project, through its events and reports, has succeeded in posing the question about geographic balance of Africa–Europe partnerships. One major tendency seems to carry over from FP7 to Horizon 2020: about 40% of all African participation comes from South Africa. Another is strong involvement of some European countries, such as Germany, France, United Kingdom and Sweden in collaborative health research projects with African countries—research institutes in these countries have previous working history and experience in sharing resources and results with partners in Africa. A corollary is the complete absence of several African and European countries in these bi-regional research projects (see Chap. 3).

Research Translation

Assessing the extent to which research outputs are translated through innovation into goods and services or new and revised policies and processes is a difficult task given the lack of validated measuring tools. Linking social, health and economic impacts to health research, investments and collaboration is all the more necessary given the considerable challenges facing health research, such as the discovery of new vaccines for HIV/AIDS, malaria and TB, or the achievement of a UHC. Although results in these fields remain fragmented, they do gradually improve health systems and healthcare services in Africa and Europe. Nevertheless, recent research development and health research programmes tend to signal positive trends regarding the measurement of progress and impacts made.

Many clinical trials address improvements and adaptations of existing treatments for specific, vulnerable target groups, such as newborns and infants, pregnant women and HIV-infected individuals, who benefit not only from the medicine, vaccine or technology being tested but also from better and more accessible preventive and curative health care. Similarly, research on neglected tropical diseases, which mainly affect populations living without adequate sanitation and in close contact with infectious vectors and livestock, is increasingly showing positive outputs. Under FP6 and FP7, several projects were funded on leishmaniasis, trypanosomiasis, schistosomiasis, Buruli ulcer, filariasis and sleeping sickness (CORDIS 2015). Results of these projects contributed to integrated diagnostic-treatment platforms and to several publications, constituting the evidence base for WHO policy revisions. This in turn contributed to the extended scope of the EDCTP programme and is also in line with the JAES.

The assessment of health research projects supported by the European Commission's (EC) Directorate General for Research and Innovation during the period 2002–2010 analysed the impact of projects on the major diseases HIV/AIDS, malaria and TB (European Commission 2011). This study confirmed the contribution of research projects to research objectives formulated by European member states and gave examples of successful projects in malaria and TB research:

- The European Malaria Graduate School, created under EVIMalaR as a follow-up to BioMalPar, has produced more than 50 European and African Ph.D. candidates in the field of malaria research. It has significantly increased the coordination of new collaborative projects between institutional laboratories within Europe and with African partners. Around 400 publications were released by the consortium's members, including a large number of high-profile publications in *Nature*, *Cell*, *Science* and so on. Due to this collaboration, Europe is now recognised as the world leader in the biology of the malaria parasite (European Commission 2011).

- TBVAC2020 is a project funded by Horizon 2020 in the field of TB. With a total budget of over 18 million euros TBVAC2020 aims at innovating and diversifying the current TB vaccine and biomarker pipeline, at setting criteria to select the most promising TB vaccine candidates, and at accelerating their development. The project builds on long-standing collaborations in previous TB vaccine and biomarker projects funded by the EC under the FP5, FP6 and FP7. TBVAC2020 involves partners from Europe, USA, Asia, Africa and Australia, many of which are global leaders in the TB field. In the global network of over 50 partners, there are four beneficiaries from South Africa and two from Senegal (Tuberculosis Vaccine Initiative 2017).

Projects funded by the FP in the field of HIV/AIDS, malaria and TB show how the strengthening of capacity through collaboration has led to greater capacity for home-grown research-based solutions to Africa's health challenges:

1. The Kenya Medical Research Institute (KEMRI) has grown to be a leading health research institution with landmark studies on impregnated bed nets and on new vaccines, having direct impact on national and international policy, and contributing to improving the lives of millions of children. Over the last 15 years there have been impressive improvements in malaria control across Africa, and in Kilifi itself cases of malaria have dropped by 90% (KEMRI 2014).

2. The Mbeya Medical Research Centre in Tanzania conducts research on the three "big" tropical diseases, and others, by evaluating new interventions, utilising vaccines, drugs and diagnostics focusing on basic research, clinical trials, epidemiological research, operational

research and social sciences. The centre has a CAP accredited research laboratory and a state-of-the-art TB laboratory (www.mmrp.org, 2017).

3. The Manhiça Health Research Centre in Mozambique has become a recognised scientific centre carrying out epidemiological and biomedical research such as a Phase II clinical trial of a TB vaccine candidate.

Long-lasting partnerships between African and European member states and research institutions seem to be a key factor for successful collaboration and continued access to funding from national and multinational programmes. All three institutions have this in common: over 20 years of continuing and intense cooperation with European countries and research institutions—Wellcome Trust and Oxford University with KEMRI, University of Munich with Mbeya Medical Research Centre and the University of Barcelona with Manhiça Health Research Centre in Mozambique. In addition to increased institutional capacities for basic research and for conducting clinical trials, African countries also benefit from the establishment of the PACTR, increased ethics capacity through the RHInno Ethics platform and through the establishment of NECs in four countries—all through EDCTP funding (see www.rhinno.net and www.researchethicsweb.org, 2017).

Strengthening national health systems is explicitly mentioned in the JAES and has been addressed by several FP7 projects. Although it was hoped that the issue of Ebola would be jointly addressed by consortia of European and African partners, only one project, the REACTION project led by the French Institute INSERM (2015), found an African cooperation partner, namely the Cheikh Anta Diop University in Dakar, Senegal. The 2014 Ebola outbreak shows the extent to which political decisions are driven by changing realities such as disease outbreaks, as the EU provided 24.4 million euros from Horizon 2020 via a fast-track procedure to supporting research projects.

Another reaction to the Ebola outbreak was an increased engagement by the European private sector in bi-regional health R&I cooperation, especially in the recent funding of Ebola projects by the Innovative Medicines Initiative (IMI), a partnership between the EU and the European pharmaceutical industry, represented by the European Federation of Pharmaceutical Industries and Associations. The total budget of the first 8 IMI-supported projects was 215 million euros, covering vaccine development and manufacture, vaccine uptake and diagnostics (IMI 2017). In view of these large and increasing amounts (especially in comparison to national health research budgets in most African countries), it is urgent to adopt a tool that will encourage compliance with existing guidelines and standards, that will indicate gaps requiring new ones to be developed and that constitutes a systematic learning platform for research partnerships.

The patenting and licensing *outside Africa* of products based on research conducted *in Africa* is, of course, a major area for future improvement. The RFI encourages research partners to make explicit statements on how they intend to address fairness in sharing of intellectual property—enabling debate, early negotiation, and gradual consensus on new standards and benchmarks.

Collaborative and multinational health research, especially between low- and high-income countries, has been a subject of controversy due to the many inequalities resulting from issues like data ownership, decision making and the application of research results in national policies and practices for capacity building (Costello and Zumla 2000; see also Chap. 7). Malawi's National Council on Science and Technology implemented a policy whereby research partners developed and enforced regulatory requirements relating to the conduct of research (National Commission for Science and Technology (Malawi) 2012). These requirements emphasise elements of fair research collaboration such as (1) the affiliation of researchers from high-income countries to local institutions, (2) the contribution to local capacity building (training, research infrastructure, technology transfer, transfer of knowledge and skills etc.) and (3) negotiating and signing appropriate MoUs or consortia agreements that aim at identifying and defining benefits of the collaborative research and a clear strategy of realising and sharing such benefits (Kachedwa 2015).

Similarly, the Council for International Organisation of Medical Sciences (CIOMS, www.cioms.ch, 2017) attempts, through a research ethics lens, to address some of the issues faced by partners in research collaborations. The CIOMS requires a sponsoring agency to ensure, ahead of the research process, that the product developed will be made reasonably available to the inhabitants of the host community or country once the testing successfully completed. However, no associated accountability mechanism exists to ensure this is done. Just as research ethics have a limited focus—mostly on participants in individual studies—so the CIOMS guidelines are inadequate for ensuring fair sharing of research benefits and therefore reduce the potential impact from sharing of intellectual property or from "spin-off" economic activity.

New research trends in African countries suggest a brighter future, however: African countries aim to increase their investments from an average of 1% of GDP (UNESCO 2010). Between 2001 and 2006, there was a 60.1% increase in medical research publications by local authors in the African region (UNESCO 2015), reflecting an increase in spending on research and/or incoming funds by African nations. Nevertheless, research expenditures and publications are only one ingredient for successful research partnerships. What makes them effective is a much bigger challenge, to which we will turn in the next section.

The Unique Potential of the RFI to Improve Research Collaborations Between Africa and Europe?

Research partnerships (or formalised research collaborations) do not apply only to high-income countries: they are not merely a luxury afforded by those with the financial means to pay for them. Research partnerships are an essential component of sustainable development of low- and middle-income countries as well. Partnerships are recognised as key to sustainable development in general (through the Sustainable Development Goal 17), while research collaborations and research networks are becoming the essential components of a strategy to deal with global or local challenges and to build national research system capacities (Nordling 2015).

However, the potential of research collaboration, partnerships and networks to build sustainable national research systems (especially in low-income countries) can only be realised if such partnerships are "fair". If all partners can derive benefits commensurate with their contributions—or perhaps even more than their contributions in the case of support for research systems in low-income countries—*and* if these benefits concern *all* aspects of the "research enterprise" and not simply sharing in a publication, then the full potential of research collaboration could be fully realised. Partners and countries (especially, again, low-income countries) should not only benefit from access to a final product or technology but also share in research system capacity strengthening and spin-off economic activities. The research enterprise is so much larger than publications: it includes the creation of jobs, increasing social capital, increasing reliability of local finance and communication facilities, sharing in intellectual property rights and the benefits deriving from these and much more.

Most, if not all, stakeholders in research are well aware of this—and many have tried and continue trying to improve the way partnerships are created and maintained, and how benefits (and costs) are shared more equitably. This applies to research collaboration between high-income countries as much as to collaborations between high- and low-income countries. The evidence-base of publications, guidelines, practical tools and even international legal instruments, like the Nagoya Protocol (United Nations 2010), is increasing (see, e.g. RFI-COHRED 2017).

The EU recently funded projects, such as TRUST, aimed at ensuring that international collaborative research using EU funding does not exploit populations in third countries (http://trust-project.eu, 2017). Similarly, the funding of the current CAAST-Net Plus project is anchored in improving policy dialogue to facilitate research collaboration between Europe and Africa in health, food security and climate change—with potential for much wider application of the project's results (https://caast-net-plus.org, 2017).

CAAST-Net Plus has been searching for ways in which the project can deliver outcomes and impact that can survive the funding limit (December 2017). In this regard, it was in 2016 that CAAST-Net Plus took the decision to adopt a partnership compliance tool under development by project partner COHRED. The RFI is a unique tool to gradually and systematically improve the way research partnerships are constructed, managed and maintained, with an emphasis on supporting low- and middle-income countries to develop their own national R&I systems.

The RFI does not invent new standards. Instead, it is a reporting tool that every major stakeholder in research should use to report on how they will behave and want partners to behave in joint research programmes. RFI Reporting Organisations (RROs) are required to provide responses to questions about the 15 most essential aspects of fairness and effectiveness in research partnerships—divided over the three phases of research collaboration: *fairness of opportunity (before)*, *fair process (during)* and *fair sharing of benefits, costs and outcome (after)*. The RFI does not ask for reports on each individual contract or partnership. It focuses on the conditions, policies and practices that RROs put in place to optimise R&I partnerships in which they are or will be involved (see http://rfi.cohred.org, 2107).

In doing so, RROs will, among other things:

- Be required to take note of existing evidence, guidelines and benchmarks, and indicate how they implement them. This makes the RFI an effective compliance tool.
- Be encouraged to identify, and then fill gaps in evidence, guidelines or benchmarking. This makes the RFI a critical learning tool.
- Be made aware of critical improvements they can make *within* their own organisation to the organisational management of research—increasing fairness, efficiency, impact and competitiveness, all at the same time. This makes the RFI an essential strategic management tool for all research stakeholders.
- Be empowering of low- and middle-income institutions and countries by enabling them to select their partners more clearly and to negotiate terms of collaboration explicitly and upfront.
- Be enabled to showcase innovations or major achievements in partnership construction and management—for which there is often no other platform inside or outside organisations. This makes the RFI an innovation tool by sharing learning.
- Be stimulated to become more transparent—to users, partners, funders and tax payers—about the social value of their institution, organisation or business. This makes the RFI a sector-specific Shared Value Report that is already being used increasingly in the private sector.
- Finally, become contributors to the first global evidence base for research collaboration and partnerships. At this time, there is no systematic evidence base—in other words, the partnership wheel is being re-invented with virtually every new partnership created, and learning ends with the end of a project. This makes the RFI a unique compliance instrument, transparency mechanism and learning platform to improve fairness, efficiency and impact of research partnerships.

Having seen the potential relevance of this tool early on, CAAST-Net Plus took a strategic decision to support its development as one of the ways in which it can make a long-term contribution to bi-regional research diplomacy and collaboration. Since then, all partners have spent time reviewing the RFI and adapting it to fit in the context of Africa-Europe research and science collaboration. Over the course of two years, the RFI will have been reviewed in and with four to five African countries—usually hosted by ministries of health and of science and technology—and in meetings involving at least six European countries, as well as major project offices in the EC. The resulting RFI is now active—institutions are beginning to conduct internal reporting—and the RFI is being reviewed for use in two major bi-regional funding calls under the Africa–Europe R&I partnership on food and nutritional security and sustainable agriculture (FNSSA).

Conclusion

There has been growing momentum in the AU–EU health research cooperation agenda, which now focuses on the infectious diseases of malaria, HIV and TB and increasingly on NDs, and on health system strengthening. Nevertheless, research partnerships between both regions need to be diversified and strengthened, while the priorities and mutual benefit of bi-regional health research cooperation partnerships must be continuously assessed. Partnerships could not only gain prominence in future programmes but also have an impact going much beyond health issues, touching on agriculture, food security, climate change and biotechnology—and these fields could be broadly integrated in research for health. Major challenges remain ahead, however. Few European businesses have yet engaged or expressed interest in engagement in bi-regional health R&I cooperation: so should initiatives such as the IMI call on Ebola be encouraged and involve African partners. Similarly, funding for cooperation in health research between Africa and Europe should not only focus on EU policy instruments and financing mechanisms but also develop new models like those used by ERAfrica and the ERA-Net co-fund for Africa on FNSSA (LEAP-AGRI). Last but not least, should the RFI become a mainstream instrument, it would provide a valuable global tool that can be used to systematically improve research collaborations involving collaborators from Africa and Europe.

Notes

1. NDs are NCDs that prevail mainly in subtropical conditions and largely affect populations living in close contact with infectious vectors and domestic animals.
2. Publications with at least one Sub-Saharan African author and another author affiliated in one of the 28 European Union member states or associate

states to the last and current Framework Programme for Research and Technological Development (FP7 and Horizon 2020 respectively) were included. In this bi-regional extract of co-publications, there are also strong co-authors from countries outside the two regions involved (e.g. Northern African countries or the United States of America).

3. The research process was first developed through a review of policies and reports on Africa–EU cooperation and health research in particular (CAAST-Net Plus 2016). Co-publications in health research from 2004 to 2015 with authors affiliated to institutions in Europe and in Sub-Saharan Africa were retrieved from Elsevier's Scopus database (www.elsevier.com, 2017). The analysis was complemented by information on EU funded health projects. Annual and evaluation reports of the FP and EDCTP were reviewed, especially in relation to the question on balanced cooperation. The principal selection criteria for a project's inclusion in the study were that (1) it involved a partnership with at least one African partner, and (2) the focus of the project was on health research. Information to address both criteria was obtained from the European Commission's website CORDIS (http://ec.europa.eu/research/, 2017) and from the Health Competence database (http://www.healthcompetence.eu, 2017). In total more than 200 FP project profiles were reviewed and 67 projects identified as relevant and grouped into six key research fields: (1) HIV/AIDS, (2) malaria, (3) tuberculosis, (4) co-infection with one of these three diseases, (5) neglected infectious diseases and (6) research on health systems.
4. The keyword sets used:

   Poverty related diseases (PRDs) (see WHO 2004): TITLE-ABS-KEY (hiv) OR TITLE-ABS-KEY (aids) OR TITLE-ABS-KEY (malaria) OR TITLE-ABS-KEY (tuberculosis) OR TITLE-ABS-KEY (dental decay) OR TITLE-ABS-KEY (diarrhoea) OR TITLE-ABS-KEY (pneumonia) OR TITLE-ABS-KEY (malnutrition)

   Neglected diseases (NDs): cf. http://www.who.int/neglected_diseases/diseases/en/ (2017) TITLE-ABS-KEY (Human African trypanosomiasis) OR TITLE-ABS-KEY (trypanosomiasis) OR TITLE-ABS-KEY (sleeping sickness) OR TITLE-ABS-KEY (Buruli ulcer) OR TITLE-ABS-KEY (Chagas disease) OR TITLE-ABS-KEY (Cysticercosis) OR TITLE-ABS-KEY (taeniasis) OR TITLE-ABS-KEY (Dengue fever) OR TITLE-ABS-KEY (Chikungunya) OR TITLE-ABS-KEY (Dracunculiasis) OR TITLE-ABS-KEY (Guinea-worm disease) OR TITLE-ABS-KEY (Echinococcosis) OR TITLE-ABS-KEY (trematodiases) OR TITLE-ABS-KEY (Leishmaniases) OR TITLE-ABS-KEY (Leprosy) OR TITLE-ABS-KEY (Hansen disease) OR TITLE-ABS-KEY (Lymphatic filariasis) OR TITLE-ABS-KEY (Onchocerciasis) OR TITLE-ABS-KEY (Rabies) OR TITLE-ABS-KEY (Snakebite) OR TITLE-ABS-KEY (Schistosomiasis) OR TITLE-ABS-KEY (Soil-transmitted helminthiasis) OR TITLE-ABS-KEY (Trachoma) OR TITLE-ABS-KEY (Yaws)

Non-communicable diseases (NCDs): cf. http://www.afro.who.int/en/clusters-a-programmes/dpc/non-communicable-diseases-managementndm/npc-features/1236-non-communicable-diseases-an-overview-of-africas-new-silent-killers.html (2017) TITLE-ABS-KEY (Cardiovascular disease) OR TITLE-ABS-KEY (Chronic obstructive pulmonary disease) OR TITLE-ABS-KEY (chronic respiratory disease) OR TITLE-ABS-KEY (Diabetes) OR TITLE-ABS-KEY (Cancer) OR TITLE-ABS-KEY (Obesity)

References

African Union & European Union. (2007). *The Africa-EU strategic partnership: A joint Africa-EU strategy*. Available from: http://www.africa-eu-partnership.org/sites/default/files/documents/eas2007_joint_strategy_en.pdf. Accessed 8 May 2017.

Breugelmans, G., Cardoso, A. L., Chataway, J., Chataway, M., Cochrane, G., Manville, C., Murali, N., & Snodgrass, J. (2014). *Africa mapping: Current state of health research on poverty-related and neglected infectious diseases in sub-Saharan Africa*. The Hague: European & Developing Countries Clinical Trials Partnership.

Breugelmans, J. G., Cardoso, A. L. V., Gurney, K. A., Makanga, M. M., Mathewson, S. B., Mgone, C. S., & Sheridan-Jones, B. R. (2015). Bibliometric assessment of European and sub-Saharan African research output on poverty-related and neglected infectious diseases from 2003 to 2011. *PLoS Neglected Tropical Diseases, 9*(8). Available from: http://journals.plos.org/plosntds/article?id=10.1371/journal.pntd.0003997. Accessed 16 May 2017.

CORDIS. (2015). *Express: Research results tackle neglected tropical diseases*. Available from: http://cordis.europa.eu/news/rcn/124183_en.html. Accessed 16 May 2017.

CORDIS. (2017). *Final report summary—HARP*. http://cordis.europa.eu/result/rcn/163257_en.html. Accessed 27 June 2017.

Costello, A., & Zumla, A. (2000). Moving to research partnerships in developing countries. *British Medical Journal, 321*, 827–829.

EDCTP. (2017). *Success stories*. Available from www.edctp.org/projects-2/success-stories. Accessed 27 June 2017.

European Commission. (2011). *Impact assessment of health research projects supported by DG research and innovation 2002–2010*. Available from: https://www.kowi.de/Portaldata/2/Resources/fp/fp-impact-assessment-health-research-2002-2010.pdf. Accessed 16 May 2017.

Innovative Medicine Initiatives. (2017). *IMI 2-Call 8*. Available from: http://www.imi.europa.eu/content/imi-2-call-8. Accessed 27 June 2017.

INSERM. (2015). *Preliminary results of the JIKI clinical trial to test the efficacy of Favipiravir in reducing mortality in individuals infected by Ebola virus in*

Guinea. Available from: http://presse.inserm.fr/en/preliminary-results-of-the-jiki-clinical-trial-to-test-the-efficacy-of-favipiravir-in-reducing-mortality-in-individuals-infected-by-ebola-virus-in-guinea/18076/. Accessed 16 May 2017.

Kachedwa, M. (2015). *Framework conditions for fair international research and innovation collaboration: Malawi perspectives.* Available from: https://caast-net-plus.org/object/news/1277/attach/M__KACHEDWA_Framework_conditions_for_fair_intl_res_and_innov_collab_MALAWI_PERSPECTIVES__.pdf. Accessed 16 May 2017.

Karolinska Institutet. (2017). *SMART2D.* http://ki.se/en/phs/smart2d. Accessed 16 May 2017.

KEMRI. (2014). *25th anniversary of the KEMRI-Welcome Trust research programme.* Available from: https://www.tropicalmedicine.ox.ac.uk/_asset/file/25th-anniversary-brochure-2.pdf. Accessed 16 May 2017.

London School of Hygiene and Tropical Medicine. (2015). Funding call: NCDs in Africa. *Centre for Global NCDs.* Available from: http://globalncds.lshtm.ac.uk/2015/05/11/funding-call-ncds-in-africa-2/. Accessed 16 May 2017.

Lozano, R., et al. (2012). Global and regional mortality from 235 causes of death for 20 age groups in 1990 and 2010: A systematic analysis for the global burden of disease study 2010. *The Lancet, 80*(9859), 2095–2128.

Ministerial Conference on Immunization in Africa. (2017). *Declaration on "Universal access to immunization as a cornerstone for health and development in Africa"*. Available from: http://immunizationinafrica2016.org/ministerial-declaration-english/. Accessed 27 June 2017.

MRC Unit the Gambia. (2017). *PROLIFICA consortium holds first meeting in the Gambia.* Available from: http://www.mrc.gm/prolifica-consortium-holds-first-meeting-in-the-gambia/. Accessed 27 June 2017.

National Commission for Science and Technology (Malawi). (2012). *National regulatory requirements and policy measures for the improvement of health research co-ordination in Malawi.* Available from: http://www.medcol.mw/comrec/wp-content/uploads/2014/07/National_Policy_Measures_and_Requirements_for_the_Improvement_of_Health_Research_Co-ordination_in_Malawi.pdf. Accessed 16 May 2017.

Nordling, L. (2015). *Research: Africa's fight for equality.* Available from: http://www.nature.com/news/research-africa-s-fight-for-equality-1.17486. Accessed 16 May 2017.

Olesen, O., & Parker, I. (2012). Health research in Africa: Getting priorities right. *Tropical Medicine and International Health, 17*(9), 1048–1052.

RAND. (2017). *Evaluating the impact of EU R&D on poverty-related and neglected diseases (PRNDs).* Available from: http://www.rand.org/randeurope/research/projects/impact-of-research-on-poverty-related-neglected-diseases.html. Accessed 27 June 2017.

RFI-COHRED. (2017). *RFI evidence-base*. Available from http://rfi.cohred.org/evidence-base. Accessed 27 June 2017.

SAnews. (2017). *SA sharpens its research quality*. Available from: http://www.sanews.gov.za/south-africa/sa-sharpens-its-research-quality. Accessed 27 June 2017.

Tuberculosis Vaccine Initiative. (2017). *TBVAC2020 project description*. Available from: http://www.tbvi.eu/for-partners/tbvac2020/tbvac2020-project-description/. Accessed 27 June 2017.

UNESCO. (2010). *Research and development: Africa is making progress despite major challenges*. Available from: http://www.unesco.org/new/en/media-services/single-view/news/research_and_development_africa_is_making_progress_despite_major_challenges/#.VwERi6R97IU. Accessed 3 Apr 2016.

UNESCO. (2015). *UNESCO science report: Toward 2030*. Paris: UNESCO.

United Nations. (2010). *Nagoya protocol on access to genetic resources and the fair and equitable sharing of benefits arising from their utilisation to the convention on biological diversity 2010, opened for signature 29 October 2010, entered into force 12 October 2014*. Available from: https://treaties.un.org/doc/Publication/MTDSG/Volume%20II/Chapter%20XXVII/XXVII-8-b.en.pdf. Accessed 16 May 2017.

WHO. (2014). *Universal health coverage in Africa: From concept to action*. Available from: http://www.who.int/health_financing/policy-framework/auc-who-2014-doc1-en.pdf. Accessed 16 May 2017.

WHO. (2016). *Director-General briefs media on outcome of Ebola emergency committee*. Available from: http://who.int/mediacentre/news/statements/2016/ihr-emergency-committee-ebola/en/. Accessed 16 May 2017.

PART III

Futures of Africa–Europe Research and Innovation Cooperation

CHAPTER 7

Towards Better Joint Work: Reflections on Partnership Effectiveness

Gerard Ralphs and Isabella E. Wagner

Abstract This chapter reflects on the issue of the "health" of cooperative projects, which it defines as the quality of the partnering relationships that underpin project-based networks. Recognising the recent proliferation of project-based networks in bi-regional research and innovation cooperation, and the challenges facing these networks, the authors propose practical applications drawing from their experiences as partners from Africa and Europe. These applications cover issues such as harmonising interests, acknowledging resources and addressing cultural specificities. They also argue that using evaluative approaches, such as partnership learning, is crucial for the partners' ability to handle success and failure.

Keywords Practitioners • Partnering processes • Partnership effectiveness • Good practices • Hidden interests • Cultural specificities • Personal identities • Power asymmetries • Informal networks • Interest-based negotiation

G. Ralphs (✉)
Human Sciences Research Council, Cape Town, South Africa

I.E. Wagner
Centre for Social Innovation, Vienna, Austria

A. Cherry et al. (eds.), *Africa-Europe Research and Innovation Cooperation*, https://doi.org/10.1007/978-3-319-69929-5_7

Introduction[1]

Africa–Europe cooperation in research and innovation can be characterised as a large and complex web of relationships, in which political, institutional and individual interests are at play. At one level, these relationships take shape in formal political agreements or project-based networks, while at another level they are realised in less formal, collaborative and interpersonal interactions, forged between professionals or institutions over time. Irrespective of their degree of formality, dynamism or purpose, these relationships are both constitutive and generative of the cooperation and are therefore essential to garnering an understanding of its present character and future potential. In this chapter, we are less concerned with the nature of the larger-scale science partnership between Africa and Europe explored in the previous chapters. Rather, we focus on the relationships between individuals and their organisations working within project-based networks, and argue that much more attention needs to be paid, by project members, their leaders and funders, to the *partnering processes* in and through which these working relationships are developed.

Admittedly, this is not a novel topic within the domains of research and innovation management and policy. In recent decades, we have seen a proliferation of toolkits to help researchers and institutions collaborating across borders to work together more effectively (see, e.g. OECD 2011; KPFE 2012). While these toolkits contain important values or principles to inform partnering processes, only recently have some of them started to provide practicable advice for how the challenges of partnering processes can be mitigated in a bi-regional context (see Chap. 6; see also, e.g. COHRED 2016). Thus, building from our understanding of the field as well as from our subjective experiences as collaborators from Africa and Europe in a multilateral network (a research and innovation cooperation support project spanning a five-year period), we offer our analysis as well as four actionable suggestions to the field of professional practice. Moreover, we suggest that more research is needed to better understand not simply what makes Africa–Europe partnerships in research *and* innovation work, but also what—looking into the future—will make them work *well.*

Why We Need Better Partnering, Now

As the previous chapters in this book have shown, collaborative research and innovation has become a more noticeable feature of Africa–Europe relations. Burgeoning collaborations operate at different speeds and are characterised by varying compositions, approaches and degrees of formality/informality. Some take form in project-based consortia, which come to an end when their funding cycles are completed. Others take form in bilateral or multilateral cooperative agreements between countries, such as the complex calculus of agreements underpinning the European Developing Countries Clinical Trials Partnership (EDCTP) or the Square Kilometre Array, which invariably involve a wide range of institutional actors and outputs and extend over many years. Others, yet, are inter-institutional arrangements specifying a broad range of continuous cooperative activities, such as student and staff exchanges or the sharing of research infrastructures; or even individual collaborative relationships that span entire careers. Many of these relationships overlap and intersect with each other and with a web of larger-scale initiatives addressing pressing challenges such as climate change, food insecurity and public health crises (European Union 2015). Whatever their catalysts, geometries or levels of engagement, partnerships are a crucial unit of analysis for understanding but also preserving the overall shape and outputs of Africa–Europe research and innovation cooperation processes.

More specifically, multilateral funding programmes such as the European Union's (EU) Framework Programme (FP) and the African, Caribbean and Pacific (ACP) Science and Technology Programme, as well as many and various national funding programmes that exist between individual African and European countries, have supported numerous research and innovation project-based networks, spanning a wide range of thematic areas (for recent mappings of multilateral projects, see European Union 2010, 2014, 2015; European Communities 2009). Typical of these project-based networks are a number of institutions or organisations that sign up to a non-returnable grant agreement, which binds a consortium to a set of contractual project deliverables that are aligned to priorities set out in a funding call (European Union 2014). Depending on their contractually defined roles and responsibilities, and their levels of in-kind contribution over and above the funders' investment, partners drawn from multiple country contexts work together in these networks at different levels of intensity and over a defined time horizon to deliver on their work plans. Given this specific bi-regional context, we use the terms "project-based network" and "partnership" interchangeably in this chapter.[2]

Studying the problems that arise when researchers and innovators from all over the world team up in project-based networks is fraught with historical, political and epistemic minefields, and can give rise to debate in which there are radically competing viewpoints. On the one hand, the language of colonialism or imperialism is sometimes invoked. For instance, in what could be considered a key paper in an emergent area of research partnership studies, Costello and Zumla (2000) describe a semi-colonial model of joint research work: "Some styles of research interaction pay little attention to ownership, sustainability and the development of national research capacity" (p. 827).

> "Postal research," whereby Western researchers request colleagues in Africa to courier to them biological samples, is still practised, though less commonly than in the past. "Parachute research," whereby researchers travel to Africa or Asia for short periods of time and take back biological samples, is still relatively common. Results of both types of research are often published with minimal representation of African or Asian input. "Annexed sites" for field research, led and managed by expatriate staff, still predominate as the model for investment. (ibid.)

On the other hand, the language of research excellence is also used to justify the selection of partners in the context of competitive funding calls. For example, in a comment to a *Research Africa* news reporter in 2009, a European national researcher, remarking on the perceived quality of research interaction between institutions and researchers from Europe and Africa, was quoted as saying that "a majority of European institutions and their scientists consider that collaborative research with Africa is 'second hand' research" (Ralphs 2009).[3]

More recently, in the health research sector, Linda Nordling, a well-known science policy journalist, has reported that the types of interaction described by Costello and Zumla seem to persist (Nordling 2015, p. 24). "Collaborations have proliferated in recent decades as international agencies have stepped up funding for health research in Africa," she writes (ibid.). "Yet African scientists say that they often feel stuck in positions such as data-collectors and laboratory technicians, with no realistic path to develop into leaders" (ibid.). In particular, Nordling investigated one very recent and widely publicised incident, in which researchers from the Kenya Medical Research Institute (KEMRI) are known to have won a court battle over allegations of institutional racism in relation to their UK counterparts in a health research partnership funded by the Wellcome Trust (see Nordling 2012, 2014a, b, 2015). There are other examples where the broader North–South partnership model has been publically brought into question by commentators (see, for instance Ishengoma 2011). However, the so-called KEMRI 6 incident is possibly a notable low for Africa–Europe research and innovation cooperation in recent decades, raising, as it does, the spectres of prejudice, structural inequality and power asymmetries.

From the above examples, it is evident that even though there are likely to be an equal number of positive experiences that could be reflected on, a few unhelpful characterisations appear to recur in the context of cooperation between researchers and innovators in Europe and Africa. One such characterisation is that Africa (often referred to in the singular, denoting a lack of diversity) is a "weak" partner, the partner in need of financial and technical support, or the partner whose capacities need to be built. The converse of this characterisation, of course, is that Europe is the financially and technically "strong" partner, the partner seeking research sites and data, and the partner who does the capacity building. Although resources differential and capacity asymmetries between African and European countries exist *de facto*, what is perhaps most telling is the extent to which these characterisations remain at odds with the political rhetoric contained in the Joint Africa–EU Strategy (JAES) (African Union & European Union 2007). The vision of the JAES is of a very different model of cooperation between the Africa and Europe, principally cooperation based on, among others, "ownership" and "joint responsibility" (African Union & European Union 2007, p. 2). In what ways could it be possible, then, for the unhelpful stereotypes to be replaced by a more constructive set of ideas and practices?

Good Practices

Studies on partnership in research and innovation have burgeoned in recent decades as collaborative models of knowledge production and technology development have emerged (The Royal Society 2011). Common to this growing body of research is an interest by scholars, funders and policymakers in the nature and outcomes of collaborative knowledge production and technology development, including its geopolitics, measurement, the cross-sectoral dimensions of collaboration, as well as the ways in which processes of working together can be improved, refined or adapted (e.g. European Union 2009, 2014; Breugelmans et al. 2015; ASSAF 2015).[4] Responding directly to some of the challenge areas for partnerships identified at the beginning of this chapter, a rich body of literature began to emerge from the mid-1990s onwards concerning the factors that influence partnership effectiveness in research and innovation projects and programmes broadly, as well as in those initiatives involving developed and developing countries specifically (Table 7.1).[5]

As shown above, several seemingly obvious aspects such as equality, joint agenda-setting and transparent communication are common to a number of these good practice frameworks. Both individually and taken together, these frameworks provide an important touchstone for prospective partners to consider in the pre-award or pre-partnership phase as well as during and after an agreement is concluded. Yet, perhaps one of the most compelling contributions to this discourse is from a 2011 OECD report, entitled *Opportunities, challenges and good practices in international research cooperation between developed and developing countries*, which argues:

> There is not, nor should there be, a universal recipe for designing and conducting research collaborations. Each situation is, to some extent, unique, and must be treated as such. Nonetheless, a variety of generic descriptive parameters can be used to characterise collaborative programmes and projects, such that intelligent choices must be made regarding their optimal value on a case-by-case basis. This process of optimisation can be viewed as a search for *balance* between various relevant requirements, not all of which can be maximised at the same time. (OECD 2011, p. 4; emphasis added)[6]

The good practice frameworks and precepts mentioned above aim to promote a set of generic partnering principles within research and innovation. In the domains of professional development practice, the emergence of capacity development organisations, such as The Partnering Initiative, which trains partnership practitioners, or the emergence of initiatives such as the Council on Health Research for Development's Research Fairness Initiative (RFI) (see Chap. 6), which is developing a reporting mechanism for R&I partnerships specifically, would suggest that the very mechanism of partnership has begun to be codified into a set of professional and organisational competencies. Indeed, many organisations working in research and innovation (universities, science councils, firms, government departments, non-profits) now have partnership offices or divisions to manage their engagement portfolios, or at least international offices for supporting partnership building. This degree of seriousness with which the partnership endeavour is being approached suggests that partnership as a modality of work and organisation requires considered and considerate organisational investment, strategy and even innovation.

Table 7.1 Key success factors for transboundary research partnerships (Hollow 2011)

Source	*Key success factors*
Gaillard (1994)	Strong mutual interest and both-sided benefits Equal involvement into the proposal and all decisions Joint decision on tools and instruments securing of their installation, maintenance and repair Include budget for a training component, if possible as part of a formal degree programme to increase commitment Salaries should be sufficient Transparency on how budget is spent Each participant organisation should include a substantial number of researchers Parties should meet regularly Communication channels must be available to secure efficient interaction between partners Scientific papers should be written jointly, with the names of the authors from both sides appearing on the published articles Collaborative programmes should be evaluated on a regular basis
KPFE (1998, 2012, 2014)	Set the agenda together Interact with stakeholders Clarify responsibilities Account to beneficiaries Promote mutual learning Enhance capacities Share data and networks Disseminate results Pool profits and merits Apply results Secure outcomes
St-Pierre and Burley (2010)—specifically refers to donor partnerships[a]	Partnership roots Positive interpersonal relations Complementarity Level of commitment Risk management Terms of engagement Governance and decision making Communication Equal footing
European Union (2014)	Equitability in all aspects (including conception, budgets, responsibilities, decision making, coordination and management) Strong leadership, coordination and management and governance Clear purpose, appropriate composition, division of responsibilities and understanding of roles Good communication, transparency and information exchange Strong interpersonal relationships and mutual trust Long-term investment

[a]ESSENCE on Health Research is a funder collaborative, consisting of donor partners working on health research for development (see http://www.who.int/tdr/partnerships/essence/members/en/, 2017)

Towards Better Joint Work: Four Key Issues for Partnership Effectiveness

In the context of a progressive and sustained interest in cooperation, our particular preoccupation in this chapter is with the issue of *effectiveness* in bi-regional partnerships/project-based networks in research and innovation. By effectiveness, we mean the extent to which the partnering activities support and enable the achievement of the goals of the partnership. As such we draw a distinction between the timely delivery of agreed project deliverables (project efficiency), on the one hand, and the "health" of the underlying network or partnership on the other hand (partnership effectiveness).[7] Our suggestion here is that these are in fact parallel processes that require different levels of management expertise and participant contribution over the life cycle of a project. As Ralphs (2013) has argued elsewhere:

> It is relatively easy for project participants to focus on the first process when evaluating their work, and there are numerous measures within project management for doing this. But success in these measures does not necessarily translate to a successful partnership. The partnership process often goes unevaluated, both because it is not predictable, and can be political. At an essential level, it comes down to difficult and likeable personalities, interaction across culture, gender and identity, institutional politics and interests, and the ability or inability to listen openly and intently—the stuff we don't often want or like to talk about.

Defined in this way, partnership effectiveness, we suggest, is a *process* and an *outcome*, requiring inputs from all partners in the stages encompassing the partnering cycle (The Partnering Initiative 2013). To state it differently, partners that sign up to a project grant agreement are not immediately effective. Effectiveness implies both a purposeful and a directed effort at *becoming effective*, which, once realised, we argue, is a desirable outcome for the partnerships itself as well as for its owners or benefactors. Reaching a "state" of partnership effectiveness, however, requires that partners work in particular ways towards their commonly defined objectives. So what are some of the key components that might help partners to become more effective on their partnership journeys? Clearly, from the above discussion, though there may be principles or factors underlying effective partnerships, there is no-one-size-fits-all approach to what constitutes an effective partnership *in all instances*. This means that the responsibility falls to partnership participants to determine what mechanisms to put in place to ensure their partnership works.

Building on the preceding discussions, we turn from a more contextual and theoretical analysis to our experiences as "African" and "European" partners in a bi-regional project-based network, CAAST-Net Plus, that was focussed on supporting research and innovation cooperation in relation to the global challenges of health, climate change and food security. At a glance, the partnership we were involved in might be seen to have met many if not all of the commonly-held criteria for an effective partnership as given in Table 7.1. This means that network participants were encouraged to conceive of and write the project together and, once approved to sign up to a non-binding consortium agreement that set out the terms of engagement between partners for the duration of the project. All of this pre-partnership documentation covered aspects such as decision making, governance, role definition and communication, and in its development, all partners were encouraged to contribute to creating an agreement fit to the particular network concerned. In addition, the network budgeted upfront for dedicated opportunities and resources for meeting in-person, for example, in the case of annual meetings, and for staff salaries to be paid for by the funder. Even though by regulation the project coordinator was required to be from an EU member state, an "Africa Region Coordinator" was appointed by the consortium to ensure appropriate voice and representation in the leadership of the network. To all intents and purposes, the network reflected the state-of-the-art in partnership theory and practice.

For all of its many positive contributions to bi-regional cooperation, what we encountered over time were four persistent challenge areas for the partnership. In the following section, we reflect on these challenges and share some suggestions for practical solutions to inspire future project-based networks working on their partnering.

Issue 1: The Problem of Hidden Interests

Some of the key reasons for engaging in research and innovation partnership activities can be summarised in at least four drivers:

- Comparative advantage: Researchers and research organisations from different countries or regions bring to a particular topic a distinct advantage, knowledge or skill, such as access to research infrastructure, populations or research sites (particularly in the context of projects with a health-related, political support or astronomical function);

- Transnational issues: Many scientific or societal problems transcend national boundaries and therefore require multi-national (or increasingly global) responses (in this sense, the notion that science "knows no borders" applies);
- Scaling up: Through working together researchers and researcher organisations or indeed countries can achieve more than by working alone;
- Getting research to market: Involving the "private sector" is essential in ensuring knowledge is translated into products or services, and even profits (Laport 2017).

Politically correct as they are (and sound), these reasons and their concepts can in fact disguise the genuine interests that partners have in a particular project: funding, reputation, data, market share and so on. Put differently, each organisation involved in a partnership, however big or small, brings to the table a set of institutional or organisational, but also individual objectives or agendas (what we refer to as "interests"). After all, in the very definition of partnership, these objectives matter centrally to the partnership's design and should of course be made explicit in the negotiation and operation of the network.[8] In our experience, however, interests that remained hidden or implicit result in a negative impact on the overall health of the cooperation.

Of course, it is difficult to know exactly the implicit interests of individuals, although in some instances these became evident in our experience. These hidden interests operated at an individual level, but also at the level of departments, institutions, organisations and whole countries. The key point to make is that we should try to be as honest as possible about implicit institutional interests and national interests, to set up projects that generate more effective partnerships because we know what to expect, and what responsibilities can be delegated. This needs to happen in the conceptualisation phase of a project.

PRACTICAL APPLICATION: Each partner should develop a statement of explicit and implicit institutional and individual interests in the conceptualisation phase of a project and an all-partner workshop on the convergence (and divergence) of partner interests in the context of the partnership's objectives should be undertaken.

Theories of interest-based negotiation suggest that it is interests (not positions) that play the defining role in a negotiation (Hamann and Boulogne 2008*). Interests are central to partnership environments, and yet can result in confusion if not handled with dexterity by the participants and managers of a partnership during its negotiation and its performance.*

Reduced to the schematic level, for private sector organisations mandated by owners or shareholders, the drivers for participating in partnerships might be underpinned by profit or growth motives, by opportunities to access new markets, by R&D opportunities, or opportunities to expand their clientele. For government organisations mandated by taxpayers, the drivers for participating in multilateral networks may be determined by very specific national policy agendas and the results of those agendas for citizens. For policy research institutes or civil society organisations conducting research to influence policy, opportunities to conduct policy-relevant research and to shape policy discourses with that research may form part of their interest in joining a partnership.

As a solution to overcome the issue of diverging interests, we would in this practical application suggest working with the following approach (or derivation thereof): During the partnership's planning phase, every partner could be encouraged to fill out a survey aiming at identifying partners' interests in the context of the envisioned project endeavour. The institutional survey responses should then be merged to a statement of institutional interests shared with all partners. The results should then support structural decisions on planning tasks and responsibilities within the consortium. In a workshop back-to-back to the kick-off meeting the statement of interests should be discussed in plenary. Awareness of the existing interests should be raised and possible solutions for situations where interests could be conflicting should be developed.

Issue 2: Personal Identities Matter in Professional Partnering Settings

Networks invariably bring professionals in research and innovation together to jointly perform specific tasks under specific sets of conditions. These professionals are however in the first instance *human beings* with particular sets of identities that matter to them or that shape their interactions with others: gender, national, cultural and so on. It is easy to gloss over these identities in a collaboration context, under the guise of an assumed common professional environment and a project work plan. We want to explicitly state that, in our experience, these identities *matter* and may play a formative role in if/how individuals interact within a particular network, on a particular task, or in a group setting. Identities may, for example, determine if individuals feel comfortable to even speak or interact in particular ways in meetings or plenary settings, or even be able to do so without fear of embarrassment or reprisal. It is essential therefore that identities are not overlooked. By creating an awareness of identity in networks collaborators can work together more honestly and openly, and different identities then can be tapped as a resource for interpersonal relationship building rather than serve as an obstacle blocking effective collaboration. As a result, there needs to be recognition of interpersonal relationships and the need to relate productively at the interpersonal level. As Gaillard reminds us: "Personal friendships among the partners are also important to overcome many frustrations in the collaboration" (Gaillard 1994, p. 57).

PRACTICAL APPLICATION: Conduct a workshop on multiple identity working contexts at the beginning of a project. Questions that can be posed could include: What do the many linguistic, gender, cultural, national, or other identities that we bring to this project mean for the effectiveness of our joint work? In what ways could these identities hinder or encourage our work?

In science diplomacy settings, including multilateral project-based networks, there seems to be pretence that all collaborators are equally educated, all globally connected, and that there is no cultural barrier. For various reasons, however, the reality is that many individuals are not able to play a role or find a position in an agreement for this very reason. Often the cultural communication aspect in partnerships is ignored, even though language translation is provided.

In this practical application, we suggest conducting a workshop where the collaborators share their experiences from previous intercultural / transdisciplinary / multi-national cooperation environments and the main dos and don'ts of their working cultures. The workshop should be moderated by an experienced trainer. The main result of such a workshop would be the emerging of understanding and therefore trust amongst the project partners to support effective team-building.

Issue 3: Money Is Not the Only Resource That Fuels a Partnership

The types of partnerships that we refer to in this chapter are project-based networks with a single major funding source. In projects supported, for example, under the EU's FP, it is obvious that funding is principally geared to European economic competitiveness objectives, that international research cooperation is one strategy to achieve this objective, and that coordinators of projects are from predominantly European organisations. Yet, in the bi-regional settings we worked in, there were many attempts made by the coordinator and the partners involved to create an equal environment for all partners. Despite these efforts, not all partners were able to assume the same amount of ownership and responsibility for project success. In our experience, for networks to function well, a number of resource pools are required in addition to finance, particularly in the kind of research support actions we performed. These are resources that all participants can bring to the table. They include in-kind contributions of time and research infrastructure, cost-saving possibilities for sub-contracting services, experience and knowledge, networks and contacts, political buy-in, strategic geographic position and many more. To develop joint ownership of the project tasks and an understanding of who is contributing what kinds of resources to project success, we suggest conducting an *ex-ante* assessment by tracking the substance of what partners bring to the table.

> ***PRACTICAL APPLICATION: Perform an assessment of all available resources at the disposal of the partnership. Brainstorm this with partners and develop a comprehensive list. Then map the resources against the partnership's goal/purpose. Is there a mismatch? Does inequality in contribution mean the partnership is unequal?***
>
> *This practical application could be done as a collaborative exercise of all partners in the planning phase of a partnership. Carefully map all the resources available to the partnership's common goal, no matter where resources originate, irrespective of the "amount" committed. Non-financial resources should be acknowledged as much as financial resources.*

Issue 4: Success and Failure Matter to "Partnership Learning"

We can think of countless examples of partnering success and failure in our own experience. By success, we mean situations where the partnering activities enabled the achievement of the objectives of the partnership. By failure we mean situations where the partnering has resulted in gaps, confusion, missed opportunities or entire breakdown between partners. The details are not important. The point we wish to emphasise here is that both successes and failures need to be seen holistically within a framework of a partnership's experiential learning (learning by doing). Where a partnership can draw on its reflective capabilities to think through these instances and integrate the learning that arises, we think that it is in, and through these situations that a network begins to encounter and experience effectiveness. Practically speaking, by documenting institutional learning processes, the loss of experience in cases of high staff turnover can be mitigated.

> ***PRACTICAL APPLICATION: Establish explicit management structures for learning from success and failure.***
>
> *There should be institutionalised processes for documenting and internally sharing lessons learned after major project milestones. The questions are relatively simple: What worked well? What didn't work at all? Why not? What should we keep, add or change? However, there are also implications for individual organisations which wish to do more partnering. These include the need to "formalise" a partnering function (Ralphs 2012), or to invest in the training of practitioners through organisations like The Partnering Initiative.*

Conclusion

Given the growing number of good practices for partners to consider in the conduct of their collaborations, this chapter serves as a reminder that that there is no one-size-fits-all formula to achieving partnership effectiveness. Partners must take an active interest in ensuring their partnership health is maintained. Using our experience and learning as partners involved in bi-regional research and innovation cooperation between Africa and Europe, we present in this chapter suggestions for improving the sense of mutual ownership and mutual respect within project-based networks, and by extension the partnership effectiveness more broadly. These can be summarised as follows: (1) individual interests should be properly identified and acknowledged in a partnership's negotiations so as not to distort the project final results; (2) cultural specificities and communication should be openly discussed and explicitly integrated into project activities; (3) financial and non-financial resources should be mapped collaboratively during the planning of the project to recognise the full contributions that partners make; (4) institutional learning should become part of management programmes and be systematically conducted so as to enable partners to reflect on the successes and/or failures of their collaboration.

Just as research and innovation partnerships and the study of research collaboration have proliferated, so too has the practice of partnering evolved into a professional competence. These developments lead to a promising opportunity, which is that partnering theory and practice could be taken into account by institutions and organisations in Europe and Africa in the pre-award or conceptualisation stage as a way of addressing some of the challenges of asymmetry and dependence. Asymmetry in resources and capacities has, in part, fed harmful narratives about Africa and Europe that pervade the partnering process, making it challenging. These challenges need to be addressed, not only at the rhetorical level but also in practice, especially in the era of the JAES, which envisions a model of partnership based on equal ownership and joint responsibility. It is our sincere hope that we have contributed usefully to how the issues that hinder partnership effectiveness can be addressed, but also that we have opened a broader and fruitful discussion among actors in the field.

Notes

1. The authors acknowledge comments on earlier drafts of this chapter received from Dr Arne Tostensen of the Research Council of Norway and Dr Andrew Cherry of the Association of Commonwealth Universities, as well as and especially the chapter's blind peer reviewers. Gerard Ralphs acknowledges Research Africa and the Human Sciences Research Council, which supported the development of this chapter.
2. To the extent that we use the terms "research" and "innovation" or "research and innovation", a qualification is also in order. In recent years, specifically with the most recent FP, Horizon 2020, the issue of innovation has been coupled to research, reflecting a desire on the part of the programme for both knowledge production and knowledge utilisation. In this context, project-based networks are supported to deliver both research and innovation activities—and as such we use the broader formulation, "research and innovation", for the purpose of the discussion.
3. To be fair to the commentator in this context, he did add the following to the statement: "This is because there is a huge lack of communication about the 'frontier research' results already obtained with African partners" (ibid).
4. For an excellent but now slightly dated literature review, see Bradley (2007).
5. For a more review on the issue of partnership evaluation commissioned by the International Development Research Centre, see Hollow (2011).
6. Three areas in which such balance must be sought, according to the OECD (2011) report, include scientific achievement and development impact, inputs and contributions from the research partners, and top-down and bottom-up approaches.

7. Gerard Ralphs is indebted to Suzanne Taylor and Lisa Burley of the International Development Research Centre for sharing the concept of partnership "health".
8. As Gaillard writes: "One of the determining conditions for successful collaboration is that the partners should be equal or at least complementary in many respects [...] [Collaboration can be successful] if the collaboration is based on a strong mutual interest and if both parties have something to gain from it" (1994, p. 57).

References

African Union-European Union. (2007). *The Africa-EU strategic partnership: A joint Africa-EU strategy*. Brussels/Addis Ababa: Africa-EU Partnership.

ASSAF. (2015). *Insights into South Africa's participation in the 7th Framework Programme for Research and Technological Development of the European Commission*. Pretoria: Academy of Science of South Africa.

Bradley, M. (2007). *North-South research partnerships: Challenges, responses and trends*. Ottawa: International Development Research Centre.

Breugelmans J. G., Cardoso A. L. V., Gurney K. A., Makanga M. M., Mathewson S. B., Mgone C. S., & Sheridan-Jones B. R. (2015). Bibliometric assessment of European and sub-Saharan African research output on poverty-related and neglected infectious diseases from 2003 to 2011. *PLoS Neglected Tropical Diseases, 9*(8). Available from: http://journals.plos.org/plosntds/article?id=10.1371/journal.pntd.0003997. Accessed 16 May 2017.

COHRED. (2016). *The research fairness initiative: Reporting for fairness in research partnerships for global health*. Geneva: Council on Health Research for Development.

Costello, A., & Zumla, A. (2000). Moving to research partnerships in developing countries. *British Medical Journal, 321*, 827–829.

European Communities. (2009). *International cooperation with Africa in FP6: Project synopses*. Luxembourg: Office for Official Publications of the European Communities.

European Union. (2009). *Drivers of international collaboration in research: Final report*. Luxembourg: Publications Office of the European Union.

European Union. (2010). *The changing face of EU-African cooperation in science and technology: Past achievements and looking ahead to the future*. Luxembourg: Publications Office of the European Union.

European Union. (2014). *Mapping of best practice regional and multi-country cooperative STI initiatives between Africa and Europe: Identification of financial mechanisms 2008–2012*. Luxembourg: Publications Office of the European Union.

European Union. (2015). *Investing in European success: EU-Africa cooperation in science, technology and innovation.* Luxembourg: Publications Office of the European Union.

Gaillard, J. F. (1994). North-South research partnership: Is collaboration possible between unequal partners? *Knowledge and Policy, 7*(2), 31–63.

Hamann, R., & Boulogne, F. (2008). Partnerships and cross-sector collaboration. In R. Hamann, S. Woolman, & C. Sprague (Eds.), *The business of sustainable development in Africa: Human rights, partnerships, alternative business models* (pp. 54–82). Pretoria: UNISA Press.

Hollow, D. (2011). *An academic review of the evaluation of partnerships in development.* Ottawa: International Development Research Centre.

Ishengoma, J. M. (2011). *North-South partnerships are not the answer.* Available from: http://www.scidev.net/global/migration/opinion/north-south-partnerships-are-not-the-answer-1.html. Accessed 26 Apr 2016.

KPFE. (1998). *Guidelines for research partnerships with developing countries: The 11 principles.* Bern: Swiss Commission for Research Partnerships with Developing Countries.

KPFE. (2012). *A guide for transboundary research partnerships: 11 principles and 7 questions* (1st ed.). Bern: Swiss Commission for Research Partnerships with Developing Countries.

KPFE. (2014). *A guide for transboundary research partnerships: 11 principles and 7 questions* (2nd ed.). Bern: Swiss Commission for Research Partnerships with Developing Countries.

Laport, G. (2017). *Time to move to an interest-driven Africa-EU political partnership.* EDCPM. Available from: http://ecdpm.org/talking-points/time-interest-driven-africa-eu-political-partnership-part-two/?utm_source=ECDPM+Newsletters+List&utm_campaign=0795947537-EMAIL_CAMPAIGN_2017_06_25&utm_medium=email&utm_term=0_f93a3dae14-0795947537-388626221. Accessed 26 June 2017.

Nordling, L. (2012). African researchers sue flagship programme for discrimination. *Nature, 487*, 17–18. Available from: http://www.nature.com/news/african-researchers-sue-flagship-programme-for-discrimination-1.10946. Accessed 16 May 2017.

Nordling, L. (2014a). *Kenyan doctors win landmark discrimination case.* Available from: http://www.nature.com/news/kenyan-doctors-win-landmark-discrimination-case-1.15594. Accessed 26 Apr [APRIL] 2016.

Nordling, L. (2014b). *Lawsuit offers lessons for alliances.* Available from: http://www.scidev.net/sub-saharan-africa/cooperation/analysis-blog/africa-analysis-alliances.html. Accessed 26 Apr [APRIL?] 2016.

Nordling, L. (2015). Research: Africa's fight for equality. *Nature, 521*, 24–25. Available from: http://www.nature.com/news/research-africa-s-fight-for-equality-1.17486. Accessed 16 May 2017.

OECD Global Science Forum. (2011). *Opportunities, challenges and good practices in international research cooperation between developed and developing countries.* Paris: Organisation Economic Co-operation and Development.

Ralphs, G. (2009). Africa-Europe research co-operation on the move. Available from: https://www.research-professional.com. Accessed 26 Apr [APRIL] 2016.

Ralphs, G. (2012). *Partnership and organisational capacity: An exploratory study of the partnering function in research organisations in East and Southern Africa* (Unpublished research report).

Ralphs, G. (2013). *Improving partnerships.* Available from: https://www.research-professional.com. Accessed 26 Apr 2016.

St-Pierre, D., & Burley, L. (2010). Factors influencing donor partnership effectiveness. *Foundation Review, 1*(4), 53–61.

The Partnering Initiative. (2013). *The partnering cycle.* London: The Partnering Initiative.

The Royal Society. (2011). *Knowledge, networks and nations: Global scientific collaboration in the 21st century.* London: The Royal Society.

Postscript | Future(s) of Africa–Europe Research and Innovation Cooperation

As discussed in the Introduction, the study of research and innovation (R&I) cooperation between Africa and Europe has been largely absent in the broader literature on bi-regional relations. This is surprising, given that science, innovation and technological fixes are the increasing focus of attention by governments and major non-state actors in addressing the range of shared global challenges, such as climate change, disease and food security. With its critical policy analysis and its profiles of a range of on-the-ground initiatives working towards addressing shared global challenges, this anthology represents a contribution towards filling that gap in the literature. Let us then, in conclusion, step back and consider the bigger picture to imagine where we have come from and where we are going.

Looking Back

In addition to sharing deep historical ties and a mixing of physical and cultural waters, Africa and Europe are geographical neighbours. Across the world, we know that neighbours, at any scale, thrive and live happier lives when they cooperate and share resources for the common good. Ensuring an ongoing conversation about our mutual challenges and interests is a fundamental prerequisite to solving them and is something we must not take for granted. As such, we should celebrate the various structures that exist to drive conversation about R&I forward. From the summit of heads of state and government that resulted in the Joint Africa–EU Strategy

A. Cherry et al. (eds.), *Africa-Europe Research and Innovation Cooperation*, https://doi.org/10.1007/978-3-319-69929-5

(JAES) framework, to the sector-specific High Level Policy Dialogue (HLPD) on science, technology and innovation (STI), and to the more frequent, often informal dialogues and cooperation between African and European academic, practitioner and policy communities, we must embrace this work with self-reflection.

Assessing the chapters gathered in this book, numerous observations can be made about the way in which African and European states, their bureaucratic structures, and the projects and programmes that they finance have evolved. The trend is clearly towards ever-greater cooperation for mutual benefit, in the context of competitive (and disruptive) economic forces that underpin local, global and digital markets. When looked at from a higher altitude, there is a stark contrast between today's world and the deeply imbalanced, exploitative and abusive structures that existed in the colonial era, where benefits to Europe were obtained at the expense of the African continent. In contrast, African and European states and private economic actors now aspire to work closely together, on an equal footing, to define and solve a broad range of societal challenges. While commentators and critics would generally not disagree in 2017 that Europe's collective economic and technological capacities exceed those of the African continent, the gap is diminishing and Africa's strong economic growth and development in recent decades will ensure that the trend towards bi-regional cooperation will continue to deepen. In any case, imbalances in capacity do not necessitate unequal partnership and collaborations; rather they simply indicate the nature of the framework conditions in which individuals are obliged to operate in, which should have no material impact on the quality of a project's outputs or the human relations that underpin these.

What we see from the analysis presented in this book is that African countries have successfully participated in several competitive funding programmes supported by Europe. Even if participation remains highly concentrated in a few countries, and especially in South Africa, the European Union (EU) strategically counts on research collaboration with African countries, as much as African countries count on their collaboration with the EU. Although the issue of capacity remains significant, it should not be seen as a binary Africa–Europe dichotomy of the "haves and have nots". The practical realities are more complex, and African countries with superior capacity can play a mentoring role to less advanced neighbours, including European neighbours and vice versa.

A key support mechanism to *diversify* African participation in the EU's R&I funding is the network of National Contact Points (NCPs), which has proved especially successful within the Framework Programme 7 (FP7) and Horizon 2020 programmes. NCPs not only offer expert and impartial guidance for those applying for funding but also assist with the identification of prospective applicants, helping them understand what these programmes offer in the local context. However, there is a consensus that more can be done to engage African and European private sector actors, to link research to commercial innovation. Here, networks such as the Enterprise Europe Network (with its Business Cooperation Centres) offer a platform to strengthen collaboration among companies and research organisations between Europe and Africa.

If, as this anthology shows, there is a genuine desire to utilise R&I to address the shared challenges we face, there is a concomitant responsibility on programme owners to focus on the achievement of outcomes—to ensure that bi-regional projects and programmes result in change, that they make a difference. While we have sought to document some of the tangible outcomes of Africa–Europe collaborations through the outcome testimonials profiled in this book, there is generally a low level of "outcome thinking" embedded in project management of STI collaborations. Indeed, both understanding and identifying outcomes are challenging, both conceptually and practically. To a large extent, we think, this reflects the ongoing preoccupation with the delivery of project outputs, that is, the specific products or services, as opposed to the difference they will make. We want to reinforce the utmost importance of considering outcomes and impact as the design stage, for any given project or programme.

Looking Ahead

In this book we have been concerned with the nature and underlying process of Africa–Europe STI cooperation, more specifically with the conditions under which cooperation takes place—the so-called framework conditions, the barriers that may hinder or improve cooperation, and the policy and programming responses that could enhance cooperation.

So what is the likely future of the specific Africa–Europe STI partnership under the JAES? Some observers, from both Africa and Europe, that we have consulted with over the years have argued that the specific Africa–Europe STI partnership is currently suffering from levels of disengagement, from mediocrity, from lack of identity and from lack of inspiration. In order to remain relevant and influence, at all levels of public policymaking and business development, the Africa–Europe STI partnership must demonstrate that it adds real and useful value to the overall Africa–Europe cooperation landscape. It has to appeal to a broader audience, engage in compelling ways with students and young people, and have an identity that is radically distinct from existing initiatives, networks and programmes.

Furthermore, the partnership should aspire to develop innovative approaches to testing and fostering cooperation, drawing in a wider set of actors, and working across a wider spectrum of value chains and within the full R&I spectrum. In doing so, it will be able to design policy and programming responses for supporting the creation and mobilisation of new knowledge, of commercial and practical value to help tackle the pressing global challenges we face as two of the world's continents. Moreover, the STI partnership cannot afford to isolate itself from the array of other sources of inspiration for innovation in new goods, services, processes and technologies. It must also recognise and work with those objectives from other domains, including some that would appear to operate in conflict with the STI agenda, such as from the domains of trade and foreign policy. The bi-regional R&I partnership on food security and sustainable agriculture, and the embryonic/emergent bi-regional R&I partnership on climate change, represent tangible opportunities in this regard.

How do we achieve these noble aims? How can we—as funders, programme managers, scientist, students, policy makers, citizens—add value? The answer is to pursue a more radical set of activities, to take significant risks and embrace the possibility of failure. For example, Africa–Europe funding mechanisms could be geared towards testing new cooperation models, not just new topics. In other words, to invest in financing models that depart from the typical non-returnable grants, towards more socially or commercially oriented spending, akin to private equity, social impact bonds or venture capital funds, albeit with a higher risk appetite. The STI partnership should aim to widen the diversity of individual partnerships, incentivising the participation of a broader array of non-traditional actors. In particular, the rhetoric about "private sector participation" must be converted into practical realities, where commercial and other private actors (including philanthropic actors from both continents) are engaged to cover the full commodity value chains and R&I spectra. Experimenting with new models will inevitably lead to some failures, but the potential benefits are also significant. Thus, risks can and must be taken, and once tried and tested, new models of STI collaboration could be rolled out and scaled up across the wider cooperation landscape. Working within the domains of STI requires, as it were, that we live our message.

Index[1]

[1]Note: Page numbers followed by "n" refers to notes.

A. Cherry et al. (eds.), *Africa-Europe Research and Innovation Cooperation*, https://doi.org/10.1007/978-3-319-69929-5

CPSIA information can be obtained
at www.ICGtesting.com
Printed in the USA
LVHW061310251020
669765LV00003B/71